T5-CFU-392

F. Cramer

Chaos and Order

Translated by D. I. Loewus

Distribution:

VCH, P.O. Box 101161, D-69451 Weinheim, Federal Republic of Germany

Switzerland: VCH, P.O. Box, CH-4020 Basel, Switzerland

United Kingdom and Ireland: VCH, 8 Wellington Court, Cambridge CB1 1HZ, United Kingdom

USA and Canada: VCH, 220 East 23rd Street, New York, NY 10010-4606, USA

Japan: VCH, Eikow Building, 10-9 Hongo 1-chome, Bunkyo-ku, Tokyo 113, Japan

ISBN 3-527-29067-2 (VCH Weinheim) ISBN 1-56081-812-3 (VCH New York)

F. Cramer

Chaos and Order

The Complex Structure of Living Systems

Foreword by I. Prigogine

Translated by
D. I. Loewus

Weinheim · New York · Basel · Cambridge · Tokyo

Friedrich Cramer
Max-Planck-Institut
für experimentelle Medizin
Hermann-Rein-Strasse 3
D-37075 Göttingen
Germany

Published jointly by
VCH Verlagsgesellschaft, Weinheim (Federal Republic of Germany)
VCH Publishers, New York, NY (USA)

Editorial Directors: Dr. Ute Anton, Dr. Thomas Mager
Assistant Editor: Eva Schweikart
Production Manager: Max Denk
Translator: David I. Loewus

Cover Illustrations: Dr. Walter C. McCrone, founder of The McCrone Research Institute and McCrone Associates, Inc., created this photomicrograph with his polarized light microscope. This fusion preparation, imaged at 100×, is of monomethyl ether of inositol, also called pinitol.

Library of Congress Card No. 93-32939

A catalogue record for this book is available from the British Library

Die Deutsche Bibliothek – CIP-Einheitsaufnahme

Cramer, Friedrich:
Chaos and order : the complex structure of living systems /
Friedrich Cramer. Foreword by I. Prigogine. Transl. by David
I. Loewus. – Weinheim ; New York ; Basel ; Cambridge ;
Tokyo : VCH, 1993
Einheitssacht.: Chaos und Ordnung ⟨engl.⟩
ISBN 3-527-29067-2 (Weinheim ...) Gb.
ISBN 1-56081-812-3 (New York) Gb.

Printed on acid-free and low chlorine paper

Composition: K+V Fotosatz GmbH, D-64743 Beerfelden
Printing: betz-druck gmbh, D-64291 Darmstadt
Bookbinding: IVB, D-64646 Heppenheim

Printed in the Federal Republic of Germany

Foreword

As is the case in great works of art, this book refers to multiple levels of reality. The world around us is full of aspects which seem contradictory: the regularity of planetary motion, the long-range unpredictability of the weather, the static structure of crystals, the dynamic and self-organizing properties of living systems. Nevertheless, we have to find a way of incorporating these various aspects into a coherent view of nature. For all who are interested in trying to obtain such an overall picture, Professor *Cramer's* book will be an excellent guide. Obviously, we are far from a final, all-encompassing answer. However, great strides have been made and we may today propose solutions to these apparent contradictions which only a few years ago would have been impossible even to formulate.

This book is first and foremost an excellent introduction to the main problems of modern molecular biology. The author's constant concern is to put these problems into the context of the progress that has recently been achieved in the theory of nonlinear systems far from equilibrium. In recent years the ideas of order and disorder have gone through a radical revision. For a long time, equilibrium structures such as crystals were considered as ideal ordered systems, while hydrodynamic flow and chemical reactions were associated with ideas of randomness and disorder. This is no longer so today. We now know that nonequilibrium can be the source of order. Self-organization is no longer outside the scope of science.

Classical dynamics was associated with time-reversible and deterministic behavior, as is indeed the case for planetary motion. But we now know that this description applies only to a restricted class of dynamic systems. The fascinating properties of unstable dynamic systems which may lead to chaos, to the generation of information, are now studied from the perspectives of an ever-increasing number of disciplines which include physics, chemistry and biology, as well as economics and social sciences. *Cramer's* presentation is clear and accurate, and will assist the interested reader in finding his way through the labyrinth of these new concepts developed in recent years.

"Chaos and Order" is the work of a man who rejects a fragmented view of the universe, who refuses to be a prisoner of preconceived doctrines. There is a feeling of excitement, a feeling that we are living at a unique moment in the history of science. The world appears to us at the same time as more strange, but also more connected and more harmonious, than to any of the generations which have preceded us. I can only hope that *Cramer's* book will inspire many readers and will receive the international acclaim it fully deserves.

Ilya Prigogine

Table of Contents

To the Reader

Dear Reader, as I sit here with the largely completed manuscript of my book before me – it can never be complete – I find myself rather surprised at the result. What do I intend to accomplish with this book? Why have I written it?

During my scientific investigations on the structure and functional processes of living things, I have been confronted again and again by difficult questions of a general nature. These questions are thought-provoking. For example: What is life? What does evolution mean? How do ideas arise in our brain? Is there a vital force? Research scientists usually ignore such questions. Indeed, they have little choice, since research can only address individual objects. Even a single cell is often too large and too complex. Therefore, a cellular component, like nucleic acid (chapter 1) or a fragment of the cell membrane (chapter 2), is isolated and examined in detail. Such investigations are so interesting and so exciting that they occupy us literally day and night. There is no time to pose deeper questions; there is no time for philosophy.

On the other hand, the results of modern biological research have had profound effects on our way of thinking and have become an integral part of our worldview. Unless we understand these effects, we risk becoming illiterates of the most general kind. Who can provide us with the required overview? Scientific results are often too complex and too difficult to grasp for even a philosopher without scientific training. Immanuel Kant was still able to deduce the philosophical consequences of the scientific revolution brought about by Newtonian physics. But it is unlikely that some thinker today will be able to fathom all the consequences of the modern scientific revolution in biology.

In this book I have sketched an outline of my thoughts on science, ideas that have gradually developed over the past decade. As the reader will immediately see, my own research results provide the scientific starting point of each chapter. I try to avoid restricting the discussion to detailed facts; instead, I try to paint a more general picture. In doing so, I have been forced to cross the traditional boundaries of scientific disciplines.

This is not intended to be a philosophical book, though philosophy will often be discussed. Nor is it a purely scientific book, though I base my reflections on scientific results. I try to do two things at once, to straddle the gap separating the "two cultures." But I do this in the hope that I might help build a bridge between science and technology on the one side and philosophy and art on the other.

Each of the nine chapters starts with a dialogue and ends with a poem. In the partly fictitious dialogues I wish to show that difficult problems need not necessarily be difficult to discuss. They can be treated dialectically through dialogue. In fact, it may even be fun to deal with problems in this way. The dialogues are intended to show, moreover, that many of the problems confronting scientists nowadays are not even all that new. Indeed, my 18th-century protagonist Georg Christoph Lichtenberg converses with people who lived as long ago as the 4th century B.C.

The book also contains difficult sections. You, my dear reader, may wish to skim over these at first. The aim and purpose of the book might even become apparent by first reading the nine dialogues. The individual chapters will then provide the supporting material for the dialogues. And those seeking to learn more can look up the relevant references in the library.

What is the purpose of the poem at the end of each chapter? The objective language of science expresses only one aspect of truth, the more informal language of a dialogue another. Poetic language is often capable of describing truths in a more valid, concise, and enduring form.

On leafing once again through my manuscript, before I give the final version to the publisher, I become somewhat melancholic. I am reluctant to part with it, since I wrote it originally for myself. I wanted to clarify my own thoughts; I enjoyed writing on these subjects and I hoped to learn more about my science and myself. The book is thus a guiding outline of sorts. Each of the nine chapters could serve as the core of a book in itself if I were to expand further on the ideas discussed therein. Someday, perhaps, I will find the time to do this.

I hope the reader enjoys these chapters. For my part, I share the wish of my Göttingen colleague Georg Christoph Lichtenberg: Whoever has two pairs of pants should sell one pair and use the money to buy this book.

Göttingen and Schloss Berlepsch,
February 1988

Fritz Cramer

Acknowledgments

I would like to thank many people. Mrs. Monika Welskop devoted much time and energy to checking, improving, revising, and again improving the manuscript, without losing an overview. Mrs. Ulrike Pruchniewicz drew many of the figures. Numerous friends in Göttingen helped me with their ideas and discussions. To name just a few: Paul Bahrdt, Otto Creutzfeldt, Manfred Eigen, Peter Richter, Albrecht Schöne, Jürgen von Stackelberg, Rudolf Vierhaus, Wolfhart Westendorf. I would like to thank all my friends, especially Wolfgang Freist and Iancu Pardowitz, who read, criticized, and improved parts of the manuscript. Finally, I would like to thank Gabriele Beck, née Cramer (1938–1988), who, over the years, discussed with me the topics covered here and provided me with many ideas. I dedicate this book to her memory.

F. C.

1. Life – A Dynamical System between Order and Decay

A conversation between Georg Christoph Lichtenberg* and Bottom** on chaos and order

BOTTOM: Good Professor, I don't understand at all. You wish to develop *a theory to explain the wrinkles in a pillow****. Truly, I had rather have the pillow itself. Without any theory or theology, whatever that means. A pillow is a fine device. Just think of everything that can be placed on it, against it, or under it. Wrinkles? Well, peradventure they may be used to determine whether Thisbe once lay there pyr-amusingly with her lover dear.

LICHTENBERG: Exactly, Bottom. An unrumpled, well-plumped, perfectly ordered pillow has no higher informational value. It is the wrinkles that lead us closer to the truth, including that of the pillow and how it is nestled in the quilt covers. From even the tiniest patterns of wrinkles *we can easily learn much about the underlying principles. The clever expedient of treating the truth itself in terms of small deviations from it* – here, the seemingly irregular wrinkles – *is the basis of the entire differential calculus and, at the same time, the reason for our weird ideas. If we regarded the deviations with philosophical rigor, the whole would often be invalid.*

BOTTOM: Professor, my fellow masters all say that I *have simply the best wit of any handicraft man* in Göttingen. Methinks it's somehow true. And now I am beginning to see why.

In our theatrical troupe we *let not him that plays the lion pare his nails, for they shall hang out for the lion's claws. Still, he is no lion's dam, just a lion fell.* The nails, then, are like the wrinkles in the pillow.

* G.C. Lichtenberg, 1742–1799, professor of physics and writer in Göttingen. Fellow of the Royal Society, London.

** Master weaver Bottom, the character in Shakespeare's play "A Midsummer Night's Dream" who directs the comic tragedy of Pyramus and Thisbe.

*** In the dialogues the italicized text is taken from Lichtenberg's "Sudelbücher" and "Briefe" and from the works of the authors mentioned.

On stage, my friend Snug is peradventure more of a lion than even a lion in the zoo, at least as far as his thoughts are concerned. Do you understand what I mean?

LICHTENBERG: Absolutely, dear Bottom. Plato himself said that the idea of a lion is more important than the physical lion itself. My colleague Kant in the Prussian city of Königsberg has lately begun to ponder such basic questions. It will be interesting to see what he concludes. *Maybe thoughts are the cause of all motion in the world.*

BOTTOM: Then the world must be chaotic indeed, Professor, because each person has his own thoughts. If all these thoughts were to affect the world, then the motions would be utterly helter-skelter. There could no longer be any order at all. As a master weaver, I am best able to judge such a state of affairs. Each fabric consists of an intricate interwoven network. If I want to design stunning new patterns, then I have to place each warp thread exactly. The weft, too, has to be considered carefully. To maintain order, I keep a plan in my head. If I didn't have this plan, could you imagine what a hopelessly tangled disorder would result. The most important thing is order. At least in weaving!

LICHTENBERG: Indeed, *ordo. Order, the active production of temporal and spatial structures. In some sciences the attempt to find a general principle, an ordo, is often just as fruitless as it would be in biology to seek a general principle or a primordial particle that could have given rise to all living things. Mother Nature does not create genera and species. She creates individuals. Our nearsightedness forces us to look for similarities in order to keep everything in focus.* The larger the reality we focus on, the less accurate the concepts of order become. Each system has its limiting degree of complexity. Perhaps a theory of wrinkles in a pillow is already beyond our conceptual scope.

BOTTOM: *That was lofty, the way you expressed it,* Professor. Again, something similar is involved in my own craft. When I weave a fabric with an intricate pattern, as I did last week for the wife of a city councilor, I can always explain the design to my assistant so he is able to weave it himself. If need be, I can even write down the steps. But only for what goes on in my own workshop, until the fabric (Lat. complexus: entwined around) is finally woven. In my workshop, everything is orderly. What the dyer does with the wool beforehand is an utter mystery to me. Sheer chaos must reign in his dye pots. At least, I can never produce the same color. And, of course, I have no idea whether afterwards the fabric is worn by ladies and gentlemen to a ball or used instead as cleaning rags. It really makes no difference to me as long as I am well paid.

LICHTENBERG: *Exactly. Order is the handmaiden of all virtues! But what leads to order?* – Well, my dear Bottom, it was a great pleasure talking with you. *No matter how you look at a thing, philosophy is always the art of distinction or separation. A craftsman like yourself uses all the phrases of the most abstract philosophy, though they are couched, hidden, contained, or latent (as the physicist or chemist would say) in commonplace maxims; the philosopher gives us clearly formulated statements.*

BOTTOM: Thank you, Professor! Then let's get to work. I at my loom and you at your desk.

LICHTENBERG: Adieu, my dear man.

Strategies of Order – The Blueprints of Life

The ever-recurring order in nature never ceases to fascinate us: the symmetry of a blossom, the geometrical pattern of a pinecone. In living things lifeless matter is ordered into highly complex structures whose schemes of organization are transmitted according to the laws of heredity. The deeper we probe into the molecular realm of biology – by studying the genetic material, the nucleic acid double helix – the clearer these schemes become. Indeed, we have even deciphered the genetic code of the double helix. This code of law ensures that order is always maintained in the realm of nature.

Order is usually regarded as something static. When we think of spatial-temporal order, we almost automatically envision an end point at which the specified order is frozen out. Crystallization is a perfect example of such order. We will see, however, that order in living systems is not a static phenomenon comparable to crystallization. On the one hand, life is the dynamical creation of order and is always accompanied by decay, the transition to chaos (see chapter 5). On the other hand, life *is* decay. The evolution of species could not be understood without a principle of selection; that is, the emergence of new species is accompanied by the extinction of others. The bacteria in a compost heap could not exist without decomposing the highly complex molecular structures of the organic matter in the pile.

Forms

Each living thing has a typical shape or form, which enables us to recognize it and to classify it in the surrounding world. Although the forms differ from one another to a greater or lesser extent, a basic pattern is always evident. In fact, nature seems to tinker with specific patterns. A leaf is always a leaf, yet it can be serrated, lanceolate, pinnate, or spiny. Its form can even be mimicked by a butterfly's wings. In other words, the same form often arises in diverse ways to serve completely different purposes. A tree leaf is optimally suited for photosynthesis; the leaflike wings of a butterfly help to camouflage it in the leaves.

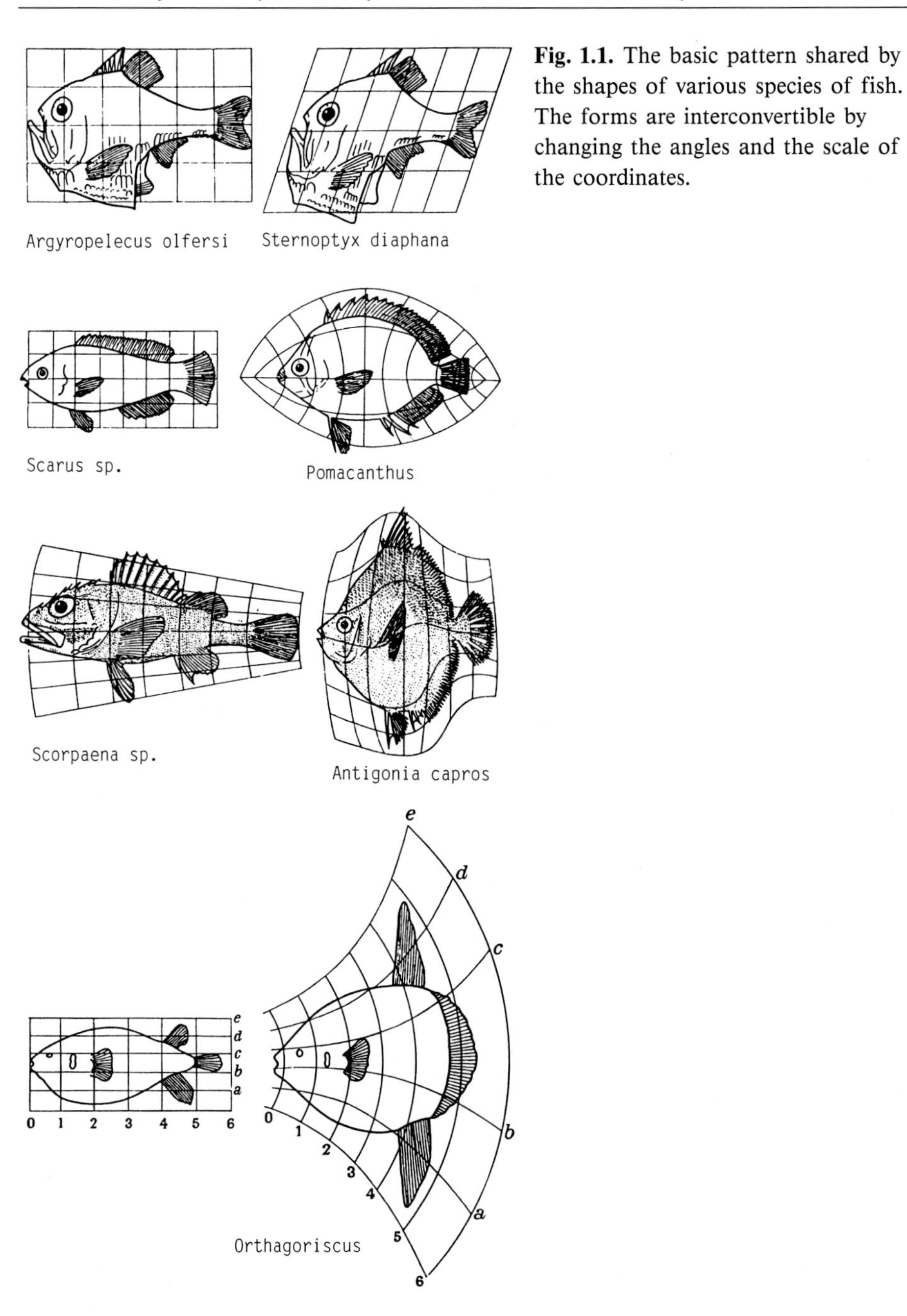

Fig. 1.1. The basic pattern shared by the shapes of various species of fish. The forms are interconvertible by changing the angles and the scale of the coordinates.

Science may be defined as the attempt to find organizing principles in a diversity of forms and, ideally, to describe these forms and their relationships to one another mathematically. Indeed, Roger Bacon called mathematics the gateway and the key to science. Numerous physical processes may be described by equations or, to be more precise, by linear differential equations. The laws of gravity (Galileo), the planetary motions (Newton), and the radiation emitted by an incandescent metal filament (Max Planck) are expressible mathematically in the form of general equations. But is this also possible for living systems?

Here, everything is much more complex. The structures and processes of living systems interact and form networks. Clearly, then, the order in living systems is much more difficult to understand. Complexity should not deter us, however. In the following pages we will try to understand the complexity of living systems and, wherever possible, to derive some general principles of nature. As we will see, it is not always straightforward to derive such principles from "basic sciences" like classical physics and astronomy or from the even more complex science of chemistry.

Many forms in the plant world are also describable by mathematical relations – for example, the spiral patterns of pinecones or composite flowers. Such systems will be examined in greater detail in chapter 6. First, by way of introduction, we will compare forms according to an approach developed essentially by Wentworth d'Arcy Thompson[1].

If the archetype of a fish, for example, is sketched in a coordinate system, then all other species of fish may be derived from it by regular transformations of the coordinates. These transformations may include changes in angles or in the scale of the coordinates as well as distortions. Each change produces a new species. In other words, the various species of fish are all derivable from a single basic pattern (Fig. 1.1).

The same coordinate transformations also allow the human skull to be compared with that of a chimpanzee. All features of a chimpanzee skull are present in a human skull. However, because the human skull has to accommodate a larger brain, the corresponding coordinates need to be adjusted. Similarly, because man no longer uses his jaw as a weapon, the coordinates of the human jawbone are smaller in scale (Fig. 1.2).

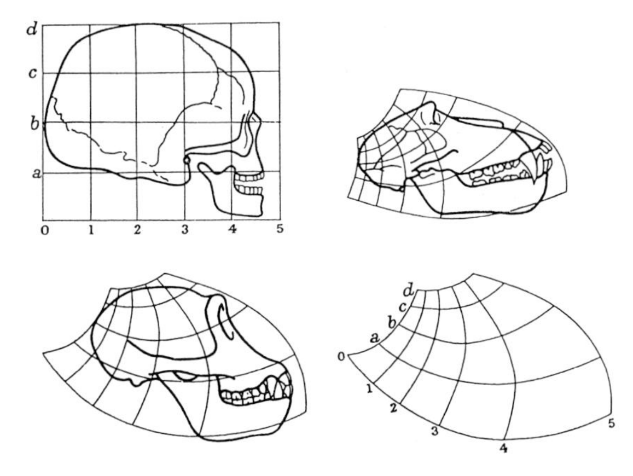

Fig. 1.2. Basic pattern shared by primate skulls. Top left: Human skull. Bottom left: Chimpanzee skull. Top right: Orangutan skull. Bottom right: Transformation of the coordinates of a human skull to accommodate a chimpanzee skull.

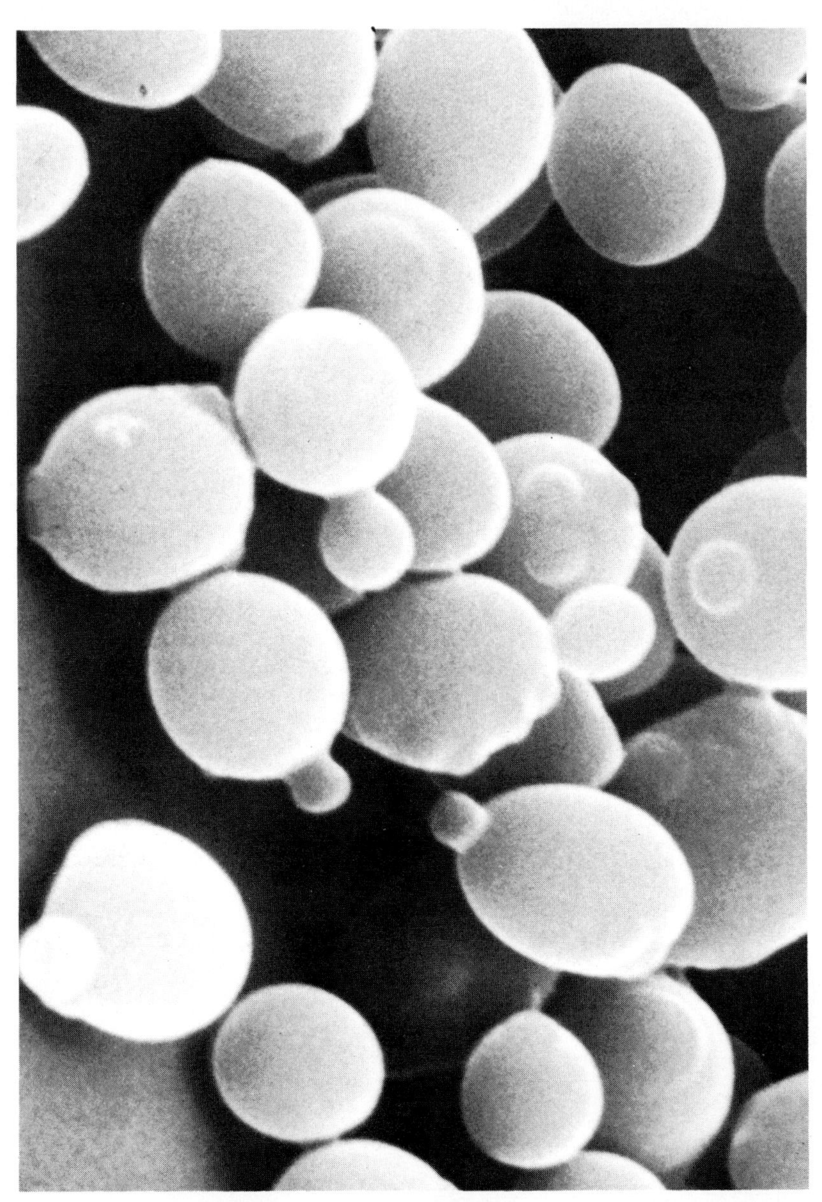

Fig. 1.3a. Scanning electron micrograph of growing yeast cells. These single-celled eukaryotes form daughter cells by budding.

In nature, then, forms are not independent and arbitrary; they are interrelated in a regular way. The blueprints of life conform to certain basic but widely variable patterns. Accordingly, the various species of fish do not exhibit completely new features such as novel limbs or organs. And even organs arising to serve new functions devel-

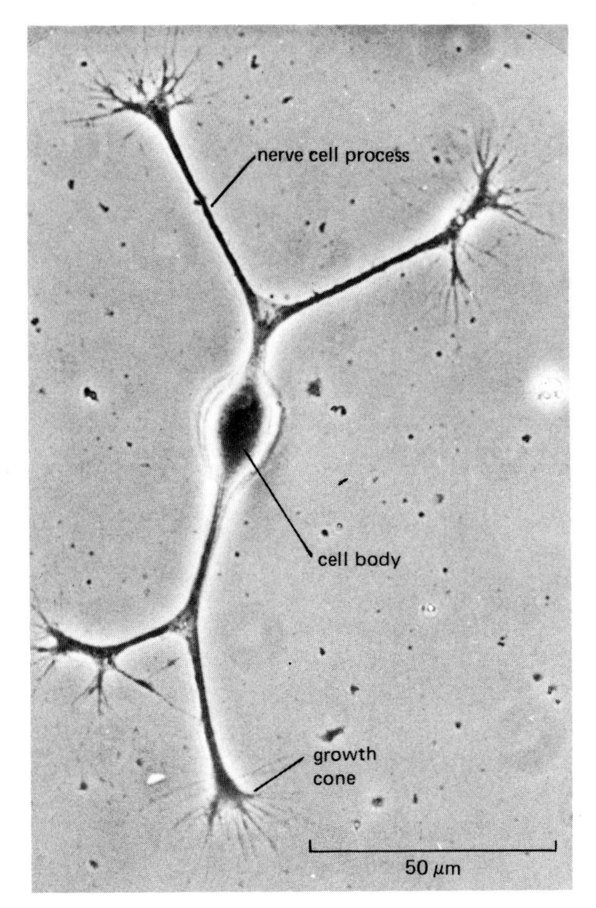

Fig 1.3b. A nerve cell viewed under a light microscope. The nerve cell, which was isolated from a chicken embryo and kept in tissue culture as nutrient medium, has started to grow long, extended processes (axons).

op according to the principle of transformation discussed here. Wings for flying, for example, evolved from the forelegs of reptiles. As Leibniz said: "Natura non facit saltus"; nature does not progress in leaps and bounds. We will see later, however, that this continuity of living systems is only apparent. At the branch points where something new emerges, disruptions of order are in fact necessary; abrupt phase changes occur. Indeed, the interplay of order and chaos constitutes the creative potential of nature.

Cells

In living systems the cell is the smallest structural unit capable of functioning independently. All animal and plant tissues are composed of cells. Most have diameters

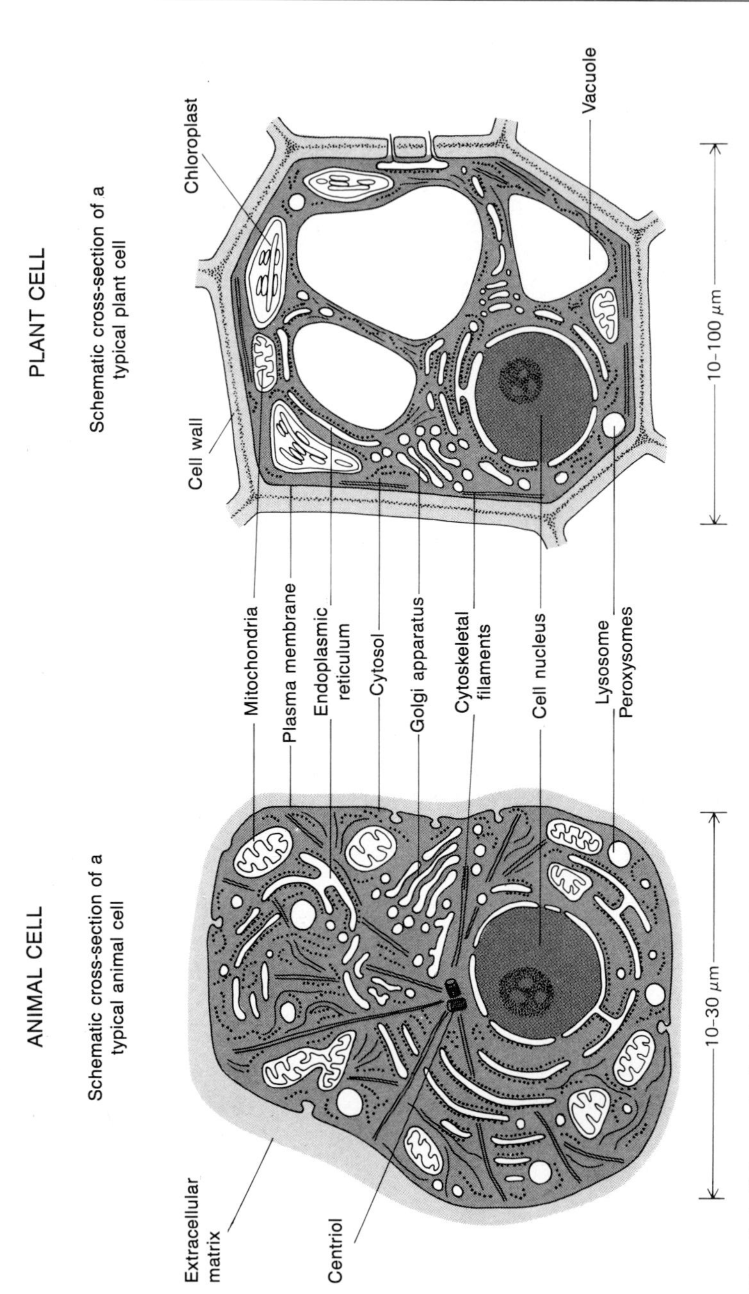

Fig. 1.4. Schematic drawing of a cell.

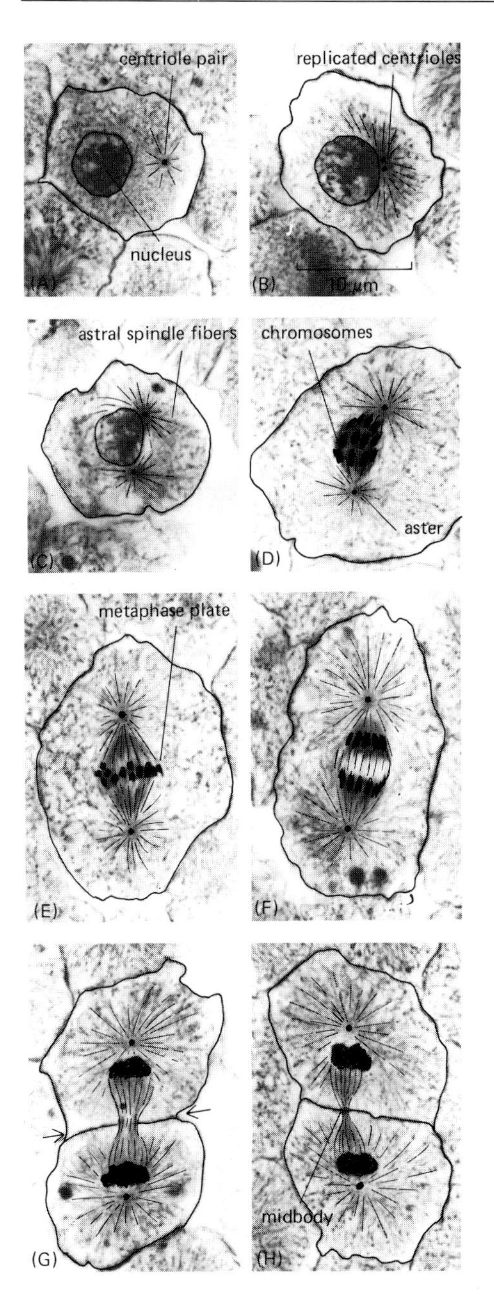

Fig. 1.5. The process of mitosis in typical animal cells (herring). (A) Interphase: The centriole pair is often small and separated somewhat from the cell nucleus. (B) Early prophase: The replicated centrioles lie near the nucleus, and fine fibers, called microtubules, radiate from them. (C) Middle prophase: The two centrioles and their radiating fibers separate along the surface of the cell nucleus to form two poles joined by so-called astral spindle fibers. (D) Prometaphase: The nuclear membrane is disrupted and the spindle fibers attach to the chromosomes. (E) Metaphase: The bipolar spindle is clearly visible; the chromosomes line up at the center of the spindle to form the metaphase plate. (F) Anaphase: The chromatids separate simultaneously and move along the spindle fibers to the poles. (G) Early Telophase: The chromatids lie at the poles. A new plasma membrane (arrows) constricts the spindle fibers remaining between the chromatids. (H) Telophase: Two new cell nuclei have formed, but they are still very compact. Cytokinesis is almost complete. The midbody or centromere still lies between the two cells.

of 1/100 to 1/1000 of a millimeter and are visible only under a microscope. Exceptions include some nerve cells, whose processes (axons) are up to one meter long, and acetabularia, a single-celled umbrella-shaped alga that can grow to lengths of 10 centimeters or more (Fig. 1.3).

Physically, cells are individuals. This means that they are able to shut themselves off from the outside world and cannot be subdivided; they exhibit individuality (Lat. individuus: indivisible). Moreover, since life exists in aqueous environments, the "mechanical fluid" of all cells is water. To isolate itself from its environment, therefore, a cell has to surround itself with a water-impermeable layer called the cell membrane, which resembles a thin film of oil. But the structure of the cell membrane is much more complicated than a simple oil film, since the cell has to take up some substances from its physical surroundings and discharge others. There are controllable entry gates or channels, receptors, signal-transducing sites, and, depending on the kind of cell, diverse recognition sites (Fig. 1.4).

Cells undergo division. Normal cell division (mitosis) starts at the cell nucleus (Fig. 1.5), the most important substructure of the cell. Such substructures, called organelles, play a role in the cell similar to that of organs in the human body. There are organelles for cell respiration (mitochondria), organelles in plants for assimilation of carbon (chloroplasts), and others. But the cell nucleus has a particularly important function. Its chromosomes contain the blueprints specifying the structure of the cell.

Although the nucleus can be removed from some types of cells, such as an egg cell, the still-living cell is no longer able to divide and produce a new cell. It goes on living until its stores of raw materials are exhausted. We will later discuss DNA (deoxyribonucleic acid), which carries the information in the chromosomes of the cell nucleus. Here, it is sufficient to note that, on cell division, the chromosomes are equally distributed between the two halves of the dividing cell, so each daughter cell receives a complete set of blueprints (Fig. 1.5). The processes involved in cell division are finely tuned and regulated to avoid errors[2].

Molecules

The most powerful electron microscopes available are capable of magnifying an object roughly 250000 times. At this magnification, details of cell organelles, microtubule structures, and even macromolecules themselves are visible. However, because of fundamental limitations – namely, the wavelength of the electron and hence its depth of penetration – it is impossible to increase this magnification to the point at which all the molecular structures of living systems are visible. Visualization of individual proteins or the double helix, for example, would require a billionfold magnification. Indirect methods, particularly chemical approaches, are therefore employed.

Chemical "magnification" consists in the strong enrichment of certain substances present in only small amounts in the cell and thus inaccessible to physical or even chemical characterization. This is accomplished by extracting large numbers, possibly billions, of cells and working up the extracts by highly refined procedures. For instance, penicillin, present in only small amounts in microorganisms, is obtainable by extraction from a whole-cell culture many cubic meters in volume. Similarly, insulin can be extracted from several thousand porcine pancreases.

Physical magnification under a light or electron microscope involves magnification of a single object. By contrast, chemical magnification no longer focuses on a single object, since this is too small for direct observation. Instead, 10^{20} copies of the pure object are prepared and the sum total of their properties is examined macroscopically. Chemistry is the art of separation, the preparation of a single kind of molecule in pure form: vitamin C – even in crystalline form – which can be used to standardize its biological effects; pure gold of high material and symbolic value; pure phalloidin (a toxin present in the fungus known as green death cap or deadly agaric), which can be used to study its effect and that of antidotes; pure growth hormone (somatostatin), which governs body growth. For example, one gram of vitamin C, provided the substance is absolutely pure, contains 10^{21} identical molecules. Although the individual molecules cannot be seen in this sum total, it is possible to determine many characteristics of the substance: its degree of acidity, its chemical formula, its medicinal value, its stability, to name but a few of those important in biochemical research. Each cell contains thousands of different molecules, which are synthesized and degraded in the cell and which make up its structural components: the cell membrane, the nuclear membrane, the respiratory chain, the chromosomes, and the DNA in the chromosomes. The relative sizes of the various cellular components are shown schematically in Figure 1.6.

René Descartes, one of the intellectual fathers of modern science, gave the following, still valid, advice about scientific inquiry in his famous treatise on the proper use of reason: "If a problem is too complex to be solved all at once, then break it up into problems that are small enough to be solved separately."[3]

This is called the Cartesian method (just as we refer to a Cartesian coordinate system). Our scientific analysis of life will employ this very method. That is, we will examine the problem first on the macroscopic anatomical level, then on the histological level under the light and electron microscopes, and finally on the chemical level.

This approach has afforded detailed insights into the structures and functions of cells, organelles, and organisms. We now understand the basic mechanisms of intermediary metabolism, cell division, and heredity (see chapter 3) and this knowledge has opened up great opportunities in medicine and pharmacology.

We have come to understand a large number of specific processes and individual substances. But do we understand the whole? An implicit assumption of the Cartesian method is that, after all the separate problems have been solved, the system can be put back together or reassembled. In other words, the sum total of the answers to the separate questions should give us one final answer. This is a central assumption. For simple systems, like those examined by classical physicists (including New-

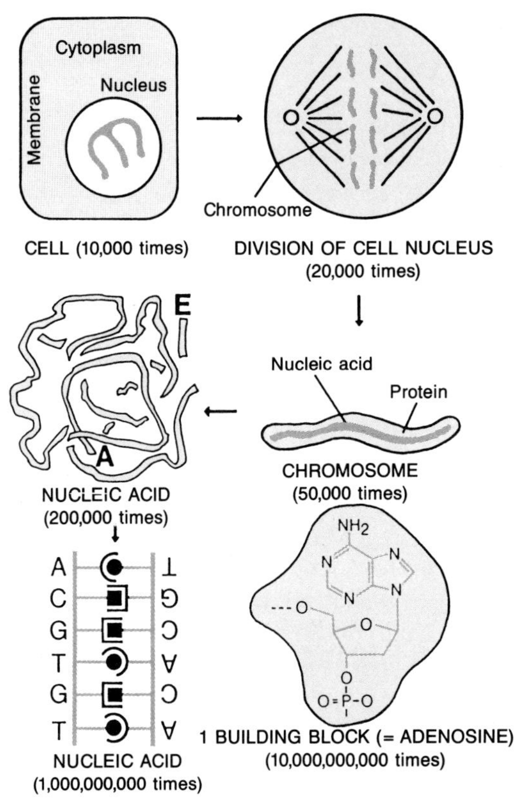

Fig. 1.6. Relative sizes of cellular components. The magnifications schematically shown in the top row may be achieved with the light microscope, those in the middle row with the electron microscope. The molecular structures in the bottom row are schematic chemical representations of physical reality.

ton), it may have been valid. Or, more precisely, the method was applied to only systems that could be reassembled.

For living systems, however, this is not the case; the whole is always more than the sum of its parts. Something is irretrievably lost when the system is taken apart – namely, life itself. Many crucial steps cannot be retraced or, to put it in technical terms, such systems are irreversible. This does not mean that the Cartesian method should not be used at all. In fact, we do not really have any choice; there is no other approach. But when the method is applied in biology, its limitations have to be kept in mind. When we dissect a living system into its parts, we examine a dead thing only. Life, a property of the whole system, is lost upon dissection. Because of the interdependence of their constituent parts, living networks cannot be dissected whenever the focus of interest is life itself, the key property of a living network.

The above considerations also hold for the cells making up a whole organism. Most cells can be isolated from native tissue and grown in culture medium. Examples include skin fibroblasts, nerve cells, blood cells, and many others. Under appropriate conditions, these cells undergo cell division and thereby multiply, eventually forming colonies. But these colonies do not constitute natural tissue. They are mere aggregates of the same kind of cell – for example, a specific type of blood cell called a leukocyte.

When the cells in a culture are all derived from a single cell, they are genetically identical. Colonies of such cells, called clones, serve as important tools in biochemistry and cell biology. Strictly speaking, they are not living things, even though they consist of living cells. Yet, as isolated parts of living systems, they can be used as models to study specific properties of life. This approach is legitimate and perfectly acceptable as long as the limitations of the model are kept clearly in mind. In this sense, biological research almost always employs models, even when the objects of study are entire living systems.

"Chemical visualization," which is obviously not possible by either light or electron microscopy, reveals that the larger cellular structures consist of molecular subunits. The macromolecules in the cell walls, in the cytoskeleton, and in the organelles and their components are packed together in a regular fashion to give higher-order structures, exactly wound to form coils, stretched to produce extended membranes, intertwined and stuck together. These structures are called tissues (see also chapter 2).

Disassembly and Reassembly

Living structures appear very complicated at first sight. They are composed of many parts and their structure-function relationships are therefore difficult to analyze. For this reason it is impossible for a scientist to understand a living thing in its entirety. At most, he or she is able to examine the overall form and draw indirect conclusions about the way of life and the behavior of an animal, a plant, or a microorganism. To understand the living system in detail, however, the object of study must be taken apart or disassembled. On a macroscopic level, this is called dissection (Greek, anatome: dissection) and has been performed since the very beginnings of modern science during the Renaissance. Taking a system apart biochemically is also an "anatomical" approach. As mentioned already, chemistry is the art of separation – the art of separating components from one another in order to characterize them exactly and describe their behavior.

The growth of an organism, however, does not involve merely the assembly of structural components, the automatic stacking of chemical building blocks according to some fixed set of instructions. The basic program involved in the design of living systems varies from case to case, depending on the particular conditions. Growth involves two interdependent processes: cell division on the one hand and the growth of individual cells on the other. Moreover, the cells in one region control the growth of those in another. One of the best-studied examples is the growth of *Caenorhabditis elegans*, a small worm discussed in greater detail in chapter 4.

Growth, then, is not a property of one part of the system alone, but a property of the system as a whole, and it is strongly subject to "feedback." For some systems, this dynamical feedback leads to chaotic states, as we will see in chapter 5.

The High Degree of Order in Living Systems Is an Extremely Improbable State

The Improbable Edifice of Life

It is a fact of everyday life that all structures sooner or later lose their order unless something is constantly done to renew it. We will later see that this is not only a sad fact of life but also a law of nature, the second law of thermodynamics. Moreover, the spontaneous emergence of order is unimaginable from our daily experience. A messy apartment with a pile of dirty dishes is just as messy the next morning, unless, of course, brownies were at work during the night. But that happens only in fairy tales. Yet living systems do display order. We ourselves are part of this order, indeed a particularly complex part with a high degree of order.

The simplest forms of life are single-celled organisms like bacteria. They are usually easy to grow in the laboratory and are therefore often used as models in biological investigations. By far the most extensively studied single-celled organism is the intestinal or enteric bacterium *Escherichia coli* (*E. coli*). Much is known about its DNA blueprint, in particular the amount of genetic information recorded there and its molecular structure. Accordingly, it is possible to calculate the number of ways these genetic plans might have been arranged in the blueprint and thus how many different ways the bacterial cell could have been constructed.

The laws of probability tell us the odds of arriving by pure chance at the correct result, one in $10^{2400000}$ (that is, 1 followed by 2.4 million zeros! Obviously, it is virtually impossible to arrive at such a blueprint by pure chance. Even an "intelligent" machine capable of checking the correctness of these countless possibilities at the rate of one per second would require far more time than the age of the universe since the Big Bang (10^{17} sec). Roughly $10^{(2400000-17)} = 10^{2399983}$ such spans of time would be necessary. The age of the universe is infinitesimal compared with the time needed for the emergence of order by mere chance. In short, life could not have arisen by a series of random events. The question of life's origin will be discussed later (chapters 4 and 7). Here, we will pose a simpler question: Is the high degree of order in living systems maintainable at all? Or, alternatively, what explains the diverse kinds of order found in living systems?

The second law of thermodynamics can be formulated briefly and trivially as follows: "Everything goes downhill." That water flows downhill is obvious. Likewise, apples fall from trees; they do not rise into the sky. Nonetheless, it took Galileo to discover the laws governing falling bodies and it took Newton to formulate later the concept of a gravitational field in which objects are heavy. We could devote much time to philosophizing about what "heavy" means and to explaining the concept of a "field," something that cannot be seen, an immaterial notion invented to explain a property, namely, heaviness. This almost sounds like a tautology. But we will leave these matters for the time being and consider them in more detail in chapter 7.

Life is not a problem of mechanics then, but primarily one of energy. How do living systems stabilize themselves energetically? How do they preserve and transmit the information necessary for their order (which, ultimately, may equally be regarded in terms of energy)?

The transfer of energy is governed by the second law of thermodynamics, developed during the last century to explain the properties of steam engines. According to the first law of thermodynamics, the different forms of energy are interconvertible. We can therefore speak of energy equivalents. For example, a performance of 1 h.p. (= 746 m^2kg/s^3, a mechanical quantity) is equivalent to a heat output of 746 J/s or 178 cal/s, which, in turn, is equivalent to an electrical output of 0.746 kW. However, the energy equivalents derived according to the first law are only possible theoretically, since the second law states that the interconversion of mechanical energy and heat cannot occur without a loss of energy.

This energy loss was first calculated by Clausius for the steam engine, which transforms heat into mechanical energy. It was formulated as a quantitative law, the second law of thermodynamics, which states that some energy is always lost in such processes. The mathematical form of the second law is as follows:

$$dS \geqq \frac{dQ}{T}$$

This equation can also be written in integrated form:

$$\Delta S = S_2 - S_1 = \int_1^2 \frac{\delta Q}{T} \text{rev} \geqq \int_1^2 \frac{\delta Q}{T}$$

Q is the amount of heat transferred and S is the entropy, which, in this equation, is a measure of the amount of energy necessarily lost. No steam engine transforms thermal energy into mechanical energy with 100 percent efficiency, regardless of how well it is insulated or constructed. Depending on the difference in temperature between the steam generated and the condensed vapor, efficiencies of 50 to 80 percent are attainable. Here, too, energy flows downhill. For a fixed amount of available energy, this law tells us that the process cannot occur in reverse. Because the transformation of energy is accompanied by energy loss according to the second law, less energy will be available upon returning to the initial state. The second thrust of a piston will be somewhat less powerful than the first. The machine slows down and eventually comes to a stop.

Entropy is not only a measure of energy loss, then, but also a measure of the irreversibility of a process. Since the flow of energy is directional in time, entropy is also a measure of time, a measure of the irreversibility of time. Although time is present in Newtonian physics, in which all motions occur in time and space, it is in principle reversible. Ideal motion does not stop. To formulate his laws, Newton had to make use of an abstraction; he distanced himself from reality. According to Newton, the motion of a body does not change in the absence of external forces to accelerate or decelerate it. In reality, though, *all* bodies are accelerated or decelerated. According to Newton, a frictionless pendulum continues to swing forever and the orbits of the planets around the sun remain practically unchanged with time. In reality, the pendulum "rubs" against the air, and our planetary system is not eternal (see chapter 5). Time is "idealized away" as a reversible, nondirectional quantity in Newtonian physics. But the second law of thermodynamics tells us that this is not valid for processes involving the transformation of energy. The decrease in the amount of available energy is thus a measure of time. "Everything goes downhill."

Hiking along a Mountain Ridge Is Possible

The discussion above would seem to imply that all living systems should collapse and that all highly ordered structures should fall apart and decay. Indeed, death is a state in which an organism is no longer supported by the influx of external energy. All its systems therefore collapse. A living organism could be defined as a system that maintains and even expands its ordered structures by constantly taking up external energy. This does not contradict the second law. A steam engine also runs for a very long time if it is continually stoked. The supply of energy is thus employed for maintaining and expanding structures. On earth, this energy comes mostly from the sun and is used by green plants for the chemical synthesis of nutrients. In other words, the electromagnetic energy of sunlight is transformed into chemical energy (e.g., glucose, starch). Our organism ingests these nutrients, adapts (digests, metabolizes) them to a form we are able to assimilate, and then transforms them into mechanical energy to produce muscular motion, into electrical energy to generate nerve impulses, into thermal energy to maintain our body temperature, or into sonic energy through the vocal chords. Fireflies even convert chemical energy into light, completing the cycle (Fig. 1.7).

The survival of life is understandable in principle, then, as long as the necessary transformations and material flows are financed by a large energy supply. In this sense life is very expensive. Once again we can draw a mechanical analogy. Systems

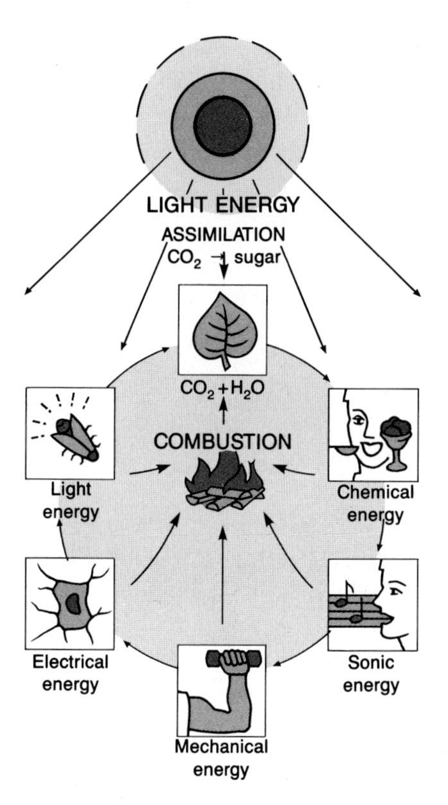

Fig. 1.7. Energy cycle in living systems. The cycle is driven ultimately by solar energy.

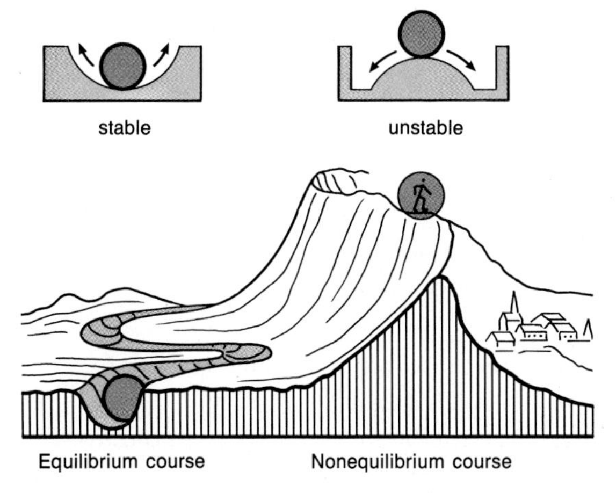

Fig. 1.8. Equilibrium systems. Left: A ball rolls until it reaches the lowest point, oscillates around this point, and finally comes to rest at this energy minimum. Similarly, a river (represented by a ball) gradually flows downhill. Right: Nonequilibrium systems behave differently. A slight shove or slip results in a precipitous fall.

low in energy are situated at an energy minimum, in a valley (Fig. 1.8, top left). Wherever possible, a river flows along the lowest part of a landscape, the river bed (Fig. 1.8., bottom left). Such systems are in equilibrium and show no tendency to change. If displaced, they return to the lowest level once again; they return to equilibrium. Life, on the other hand, is more like mountain climbing or, better, hiking along a mountain ridge. Once on top, a hiker might follow a narrow, dangerous ridge, provided he is careful, intelligent, and somewhat skilled in mountain climbing. If he becomes exhausted, though, he is in great danger; he could stumble, fall into a chasm, and perish. That's life!

Dissipative Structures – The Generation of Forms by Consumption of Energy

Living organisms need to consume energy, even in hibernation. This energy allows them to form structures that are far from equilibrium. According to Ilya Prigogine, these are called dissipative structures; that is, they consume energy (dissipate, from Lat. dissipare: distribute)[4].

The physical principle underlying dissipative structures is easy to demonstrate for simple systems. A well-known example is what is referred to as Bénard instability, which arises during thermal convection in a liquid. If a liquid is placed between two plates, the lower one of which is at a higher temperature, the liquid will warm up and heat will be transported upwards. When the energy difference between the lower and upper parts of the liquid attains a certain value, wavelike patterns set in (Fig. 1.9). This is an example in which the consumption of energy spontaneously gives rise to order. If the temperature between the plates becomes too large, however, turbulence and disorder result. We can observe Bénard patterns on the surface of a cup of hot coffee, where the condensed clouds of vapor often form highly regular patterns. A chemical example of a dissipative structure is provided by the Belousov-Zhabotinskii reaction, the oxidation of an organic acid (malonic acid) with potassium bromate in the presence of a catalyst. This remarkable reaction proceeds discontinuously in homogeneous solution and a temporal-spatial pattern is thereby formed (Fig. 1.10).

Benno Hess and co-workers recently presented a very beautiful example of a moving wave of chemical activity in the Belousov-Zhabotinskii reaction[5]. A number of biochemical reactions also result in pattern formation under certain conditions. The degradation of glucose by the corresponding cellular catalysts displays a pattern very similar to the Bénard waves observed for thermal convection[6].

Under certain conditions, the degradation of glucose spontaneously gives rise to a temporal order in the form of oscillations that might function as a kind of biological clock[7,8] (cf. Fig. 1.11). The oscillation frequency is dependent on the concentration of the components.

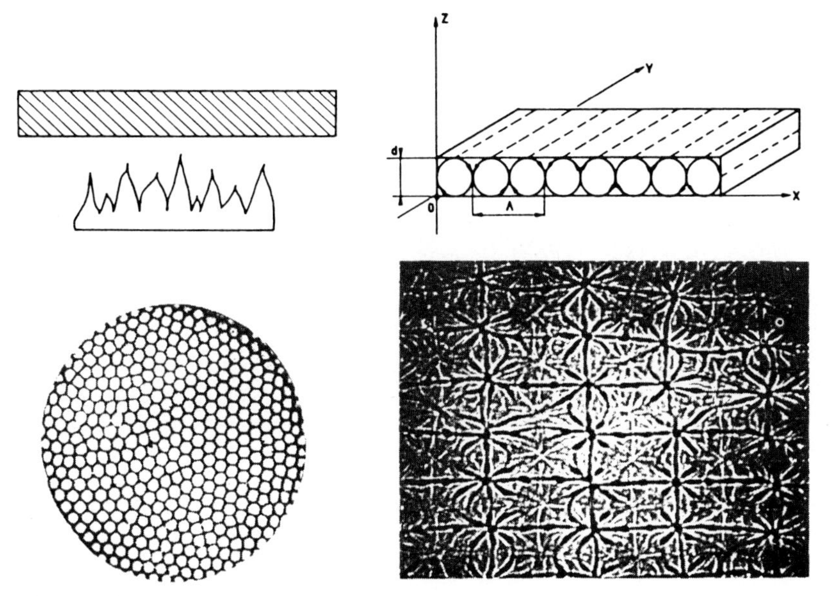

Fig. 1.9. Spontaneous generation of patterns (Bénard waves) in a heated liquid. Top left: A layer of liquid is heated from below. As long as the temperature difference between the lower and upper surface is small, heat is transferred smoothly. Top right: For larger temperature differences, convection is observed and Bénard waves form. Bottom left: The cell-like structure of Bénard instability viewed from above. Bottom right: Pattern formation at higher temperatures.

Fig. 1.10. Spiraling waves in the Belousov-Zhabotinskii reaction[5].

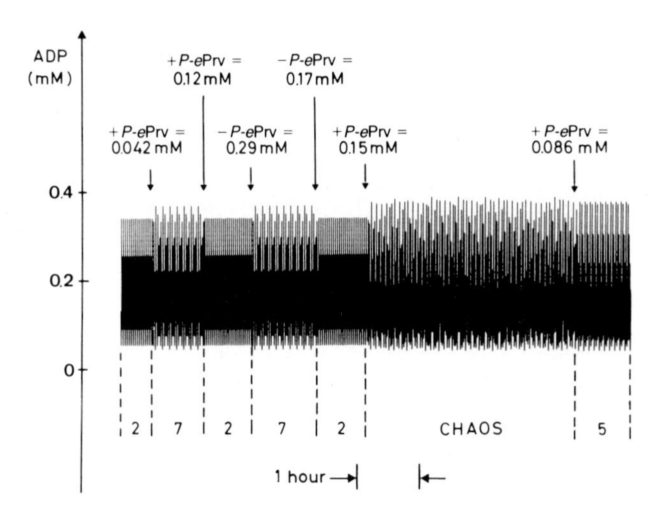

Fig. 1.11. The oscillation of an enzymatic reaction involving two enzymes (pyruvate kinase and phosphofructokinase) competing for the same substrates. Small changes in the concentrations can lead to an increase[7] in the oscillation frequencies and amplitudes or cause them to become chaotic as described by the Verhulst law (see chapter 6).

Cycles – Everything Flows

Where do cells or higher organisms obtain the energy to maintain an improbable state that is far from equilibrium, a dissipative structure? For this purpose living organisms have in common a large number of biochemical mechanisms. Particularly noteworthy are the processes that degrade glucose or, as it were, "burn" dextrose. In these circular processes, called cycles, glucose is converted into other compounds, which serve in turn as new building blocks or undergo complete combustion to yield carbon dioxide and water. One example is the citric acid or citrate cycle (Fig. 1.12) in which citric acid, derived from glucose, is ultimately degraded to CO_2 and water with production of energy. The cycle shown in Figure 1.12 is only one of many similar metabolic cycles elucidated by biochemists. These cycles provide the energy needed to maintain living systems far from equilibrium. Analogous and often much more complicated cycles are found in all areas of biochemistry. Such systems are regulated by feedback and, like true mechanical oscillators, they sometimes start to oscillate when perturbed.

Heredity – The Material Flow of Tradition

One of the characteristic properties of life is heredity. The order of living things is preserved, at least on a human time scale. Even the most peculiar features, such as the famous Habsburg lip, are inherited.

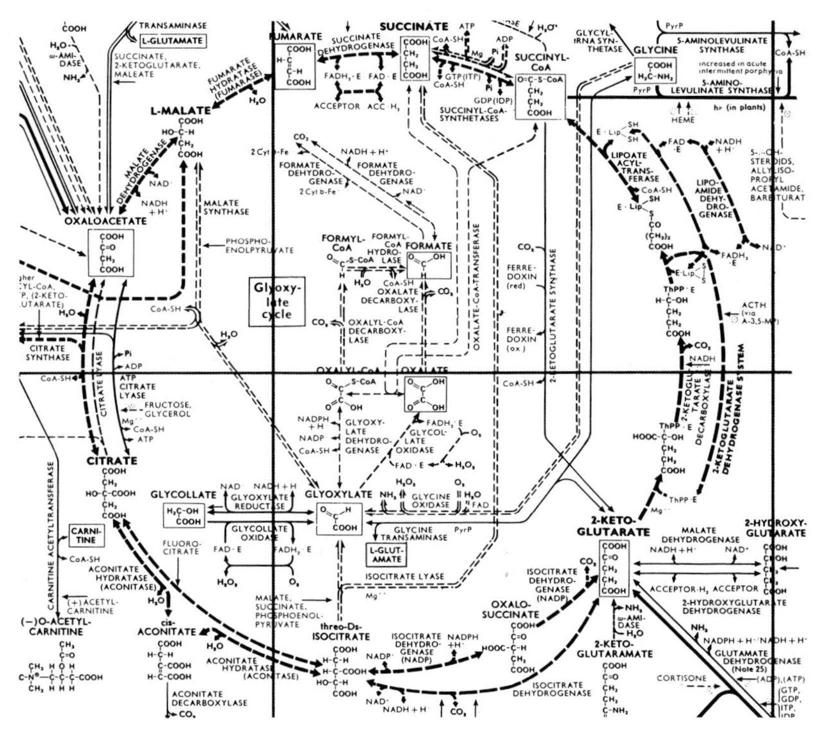

Fig. 1.12. The citrate cycle: it provides a large part of the energy required by cells.

Gregor Mendel was the first to formulate once and for all the laws governing heredity. He accomplished this by crossing different pure strains of pea plants and observing the number of times specific traits appeared in the progeny. He found that, though some traits are often latently present, they are still by and large indivisible; they are "atoms of inheritance." When pea plants with white blossoms and those with red blossoms are crossed, the first generation consists of pink progeny, displaying what appears to be a homogeneous mixture of the parental traits. When these progeny are crossed, the next generation has red, white, and pink blossoms. Accordingly, all hereditary traits are present in two copies and are latently present. For example, when both copies specify the trait "red," (A,A), the blossoms are deep red. When neither copy specifies this trait, (a,a)*, the blossoms are white. For the intermediate case, (A,a), the blossoms are pink. Therefore, the first generation (F_1) necessarily consists solely of pink progeny, (A,a), since this is the only mixture possible. The situation is different, though, when the F_1 plants are crossed. A simple statistical anal-

* In genetics, a small letter indicates the absence of the specified trait: a = A is not present, y = Y (yellow) is not present, r = R (round) is not present.

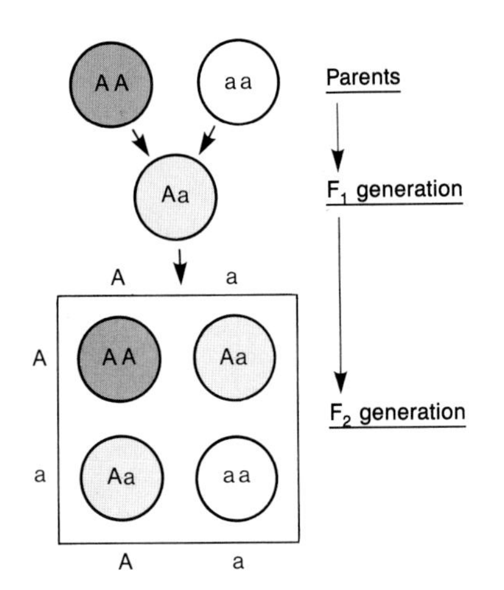

Fig. 1.13. Schematic depiction of the first Mendelian law. Inheritance of blossom color in pea plants. The parents are pure red and pure white, their progeny pink. In the second generation, the pure traits reappear.

ysis reveals that, in the F_2 generation, (A,A), (A,a and a,A), and (a,a) are formed in a ratio of 1:2:1. This is the first Mendelian law (Fig. 1.13).

Mendel further found that when two different traits are present, they can be inherited independently of each other. For instance, peas are either smooth or wrinkled, green or yellow.

The smooth and yellow traits are dominant; that is, they are expressed preferentially. The green and wrinkled traits are recessive; they can be latently present. When a plant with smooth yellow peas and one with green wrinkled peas are crossed, the first generation produces only smooth yellow peas. When these plants are crossed, in turn, three-fourths of their progeny produce yellow peas and one-fourth green peas; independently, three-fourths of the peas are smooth and one-fourth wrinkled. Therefore, both smooth and wrinkled green peas are also found. These combinations of traits were not present in the parents (Fig. 1.14).

The only way to explain these relationships is to assume an "atomic" unit of inheritance, which is passed on either completely or not at all to the progeny. This unit is called a gene (Greek: γεναιειν, *genaien:* generate). Accordingly, the science of heredity is called genetics.

Things are often somewhat more complicated when, say, a gene is associated with a chromosome present in just one copy. This is the case for hemophilia, a disease caused by the inability of the body to produce one of the clotting factors. The gene for this clotting factor is located on the X chromosome, present in only one copy in males. Because of the double set of chromosomes in higher organisms, particularly human beings, the female might carry the defective gene latently but still use a copy of the healthy gene on the second X chromosome to produce a functional clotting factor. Although they are capable of transmitting the disease genetically, therefore, females themselves do not suffer from it. On the other hand, males have only one X chromosome and hence are not protected from the disease if they carry the

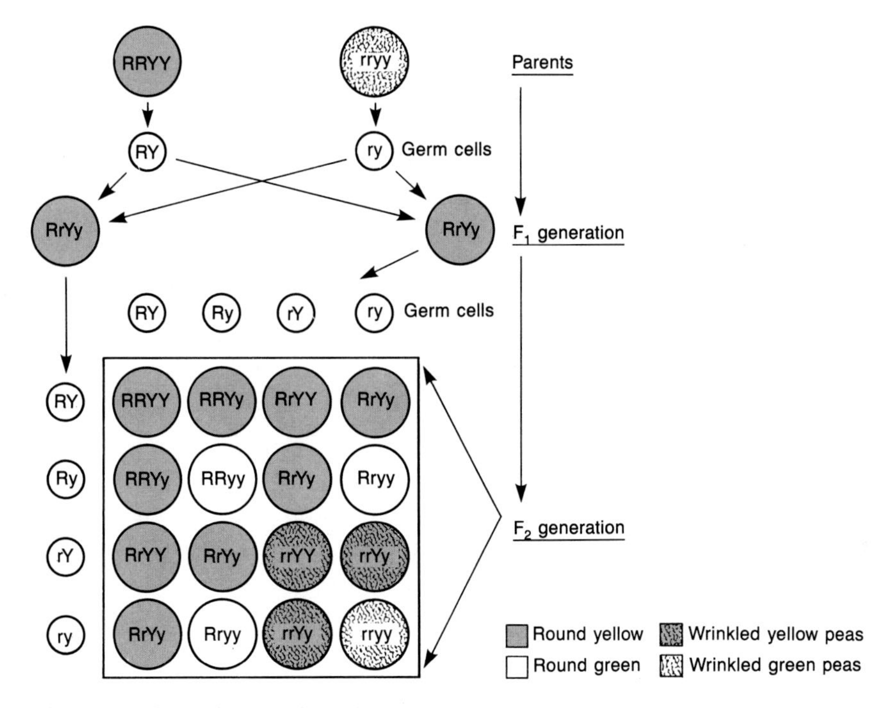

Fig. 1.14. Schematic depiction of the second Mendelian law. Yellow and round peas are dominant and these traits are inherited independently of each other. Accordingly, in the F_2 generation, three-fourths of the peas are round and – independently – three-fourths of the peas are yellow.

defective gene. They are haploid with respect to this chromosome. Hemophilia appeared in the ruling houses of Europe at the end of the 19th century. An exact genetic analysis of the related families within the European dynasties allows the course of inheritance to be traced. At least in these families, the hemophilia originated in the ovaries of Queen Victoria through an error, a mutation. This does not mean, of course, that the disease did not occur in other families. Because of the prominence of the people involved, however, the inheritance in this case has been documented particularly well. The respectable, highly moral, austere queen, the symbolic figure of the Victorian era, was thus the origin of a sign of decay. Life is order and decay.

The Double Helix – Molecular Information

The logical consequence of the Mendelian laws is that specific traits are passed on either in their entirety or not at all, that is, as a unit, as specific genes. The Mendelian laws may thus be regarded as an atomic theory of heredity. Since the Greek antiquity and Democritus, the indivisible units making up all things have been referred to as atoms. A-tomos (Greek: ατομοσ) means indivisible.

The gene was at first a purely conceptual notion, like the physical atom. Modern nuclear physics has shown, however, that the atom is not the smallest indivisible particle of matter. Radioactive decay of the atomic nucleus can occur and nuclear fission is possible; there are nuclear fragments, positrons, neutrinos, quarks, and so forth. Nonetheless, the basic concept of Democritus is still valid. For the world in which we live – its elements and chemical compounds, its numerous biochemical processes – the atom is still the smallest physical unit of matter. But what happened to the concept of a gene? Before answering this question, I wish to make a short historical digression. After purely biological studies had revealed that the carrier of heredity was localized in the chromosomes, scientists tried to isolate a substance from the chromosomes that might correspond to a gene or collection of genes. This is the usual pathway followed in scientific research. First, a concept that has little or no physical and experimental support is formulated, then the material correlate is sought. The concept of atoms was proposed 2400 years ago but not until the 20th century were they actually identified. Later, the concept was shown to be inaccurate and had to be revised. In reality, atoms are divisible.

We will later see that for genes the situation is very similar; the basic genetic units are further subdividable into introns, exons, promoter regions, triplets, bases, and so forth. But this does not change the fact that the concept of a gene, like that of an atom, is still valid. An organizational scheme is always subject to analysis and revision or, as it were, disassembly and reassembly. Order and decay go hand in hand; decay is the logical counterpart of order. One of these cannot be conceived of without the other.

Let us return to the history of the gene, however. In 1944 pneumococci (the causative agents of pneumonia) were discovered to transmit their genetic information solely via nucleic acid. Nucleic acid is the carrier of this information, therefore, and a gene must correspond to a segment of the long nucleic acid strand. This realization stimulated scientists to isolate these macromolecules in pure form and to elucidate their structures. I, too, was involved in this research to a small extent in 1953, when I joined the laboratory of Alexander Todd in Cambridge, England, and began work on the synthesis of nucleic acid building blocks. While I was there, two biophysicists, rather odd fellows, came over from the neighboring Cavendish Laboratory to ask us some questions about the chemistry of nucleic acids, which were a complete puzzle to them. They told us that they intended to elucidate the three-dimensional spatial structure of these molecules. This challenging, but presumptuous goal amused us at the time, since in our laboratory we were just beginning to understand the rudiments of the one-dimensional structure of these complex molecules. We gave them a few suggestions, but did not really take them very seriously. Several months later the famous seminar on the structure of the double helix took place. The two "odd fellows"

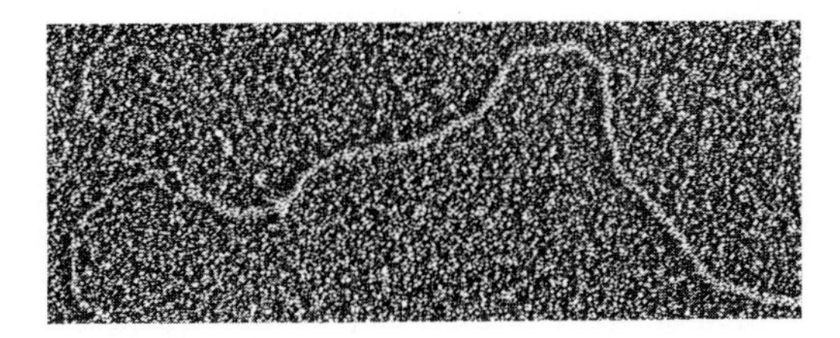

Fig. 1.15. Nucleic acid magnified 100000 times.

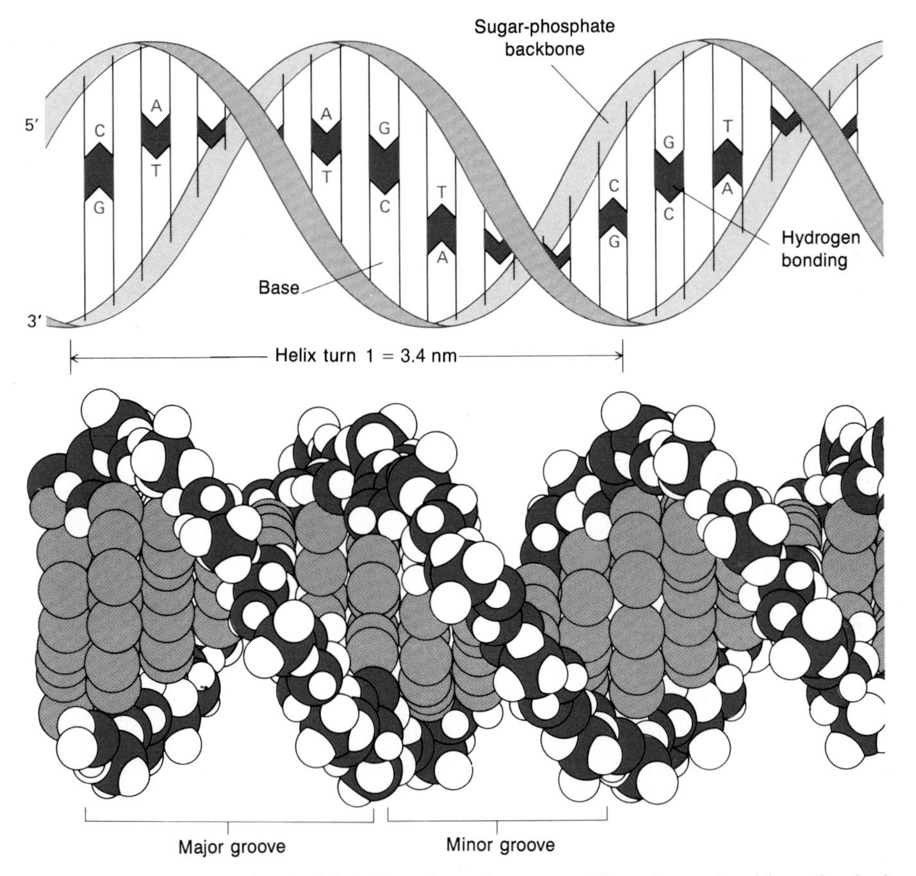

Fig. 1.16. A segment of a double helix enlarged twenty million times. On this scale, the human genome (the entire set of genes) would be about 50000 km long.

turned out to be Jim Watson and Francis Crick. The story of this momentous discovery is recounted in the bestseller "The Double Helix."[19]

The isolation and chemical synthesis of pure nucleic acids is relatively simple. It is even easy to show visually that these molecules are indeed huge, resembling a long chain. Pure, uncleaved nucleic acids form long threads. These threadlike structures

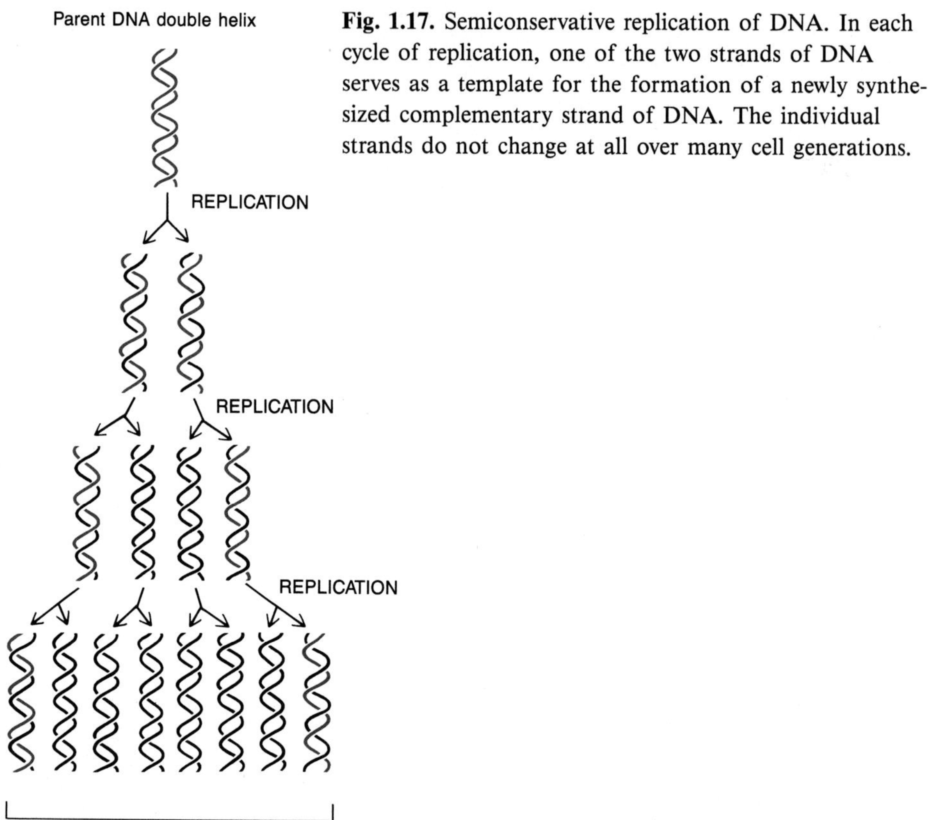

Fig. 1.17. Semiconservative replication of DNA. In each cycle of replication, one of the two strands of DNA serves as a template for the formation of a newly synthesized complementary strand of DNA. The individual strands do not change at all over many cell generations.

are even visible to the naked eye, although the slimy fibers actually observed contain not just one but billions of molecules. These nucleic acid threads can be easily withdrawn from solution and wound around a glass rod.

Threads corresponding to a single nucleic acid are visible, just barely, under the electron microscope (Fig. 1.15). These nucleic acid strands are about 2.5×10^{-6} millimeters in diameter and, in man, about 990 millimeters long. Their ratio of length to diameter is therefore one billion to 1. They are indeed *tapes* of encoded information. If the DNA strand were as thick as a normal thread (about 1 mm), it would be 1000 kilometers long.

Imprinted on this tape of information are the chemical characters A, G, T, and C, arranged in such a way that they constitute a kind of writing or code. But this code can be represented through formulas only, since the chemical characters themselves are too small to be seen with the naked eye (Fig. 1.16). With chemical magnification, however, we can "read" the nucleic acid code. This method is described in more detail in chapter 3, where the structure of a complete gene is also shown (Figs. 3.3 and 3.4).

Inherent in the double-helical structure of DNA is a possible mechanism of replication during cell division. Indeed, an essential requirement for the storage of genetic information is that it can be copied accurately, so the two daughter cells contain identical genetic information. The double-helix structure ensures that this requirement is met (Fig. 1.17).

During replication, the double helix unwinds and a complementary strand is formed synchronously along each of its two strands, affording two new double strands identical to the parent strand. In this way, the molecular library of one thousand or so volumes is accurately copied during cell division (20 to 80 minutes). Accurately? We will later see that the few errors occurring during this process are crucial for the dynamics of living systems. The minimal inaccuracy of the copying process – less than one error is made each time the entire library is copied – makes possible both the variability of species and evolution. The error has a positive effect on the overall dynamics of life. Life is order and decay.

Proteins – Costly Precision Work

An organism does not consist of DNA alone. DNA provides only the information, the instructions, the master blueprint for an organism. Just as an audiocassette does not produce music all by itself, DNA alone does not constitute a living structure. Translation of the blueprint into an actual structure or "work" of life is like translation of an audiocassette into a work of music; a decoding mechanism, a synthesizer, and an amplifier are required. For some steps of protein biosynthesis these mechanisms have been elucidated in great detail. Proteins are composed of twenty amino acids, which have to be exactly linked in a prespecified way to form the correct protein (p. 28, top). As we have seen, there are countless ways of constructing a macromolecule starting from twenty building blocks. But only certain combinations make sense, that is, result in functional proteins that the cell is capable of using and incorporating. An example is the pancreatic hormone insulin. Only the specific sequence of amino acids corresponding to insulin has an effect on the blood sugar level (p. 28, bottom).

The human organism contains more than 10000 different proteins (not counting the huge number of diverse antibodies). All are coded for by DNA, often in several copies. These 10000 or more proteins, in turn, have to be correctly situated in the living network, so they are able to perform their functional roles in the overall processes taking place therein. The information required to specify and direct this organization must also be furnished by the DNA to a large extent, in fact usually in fine detail, since we are dealing with an organizational hierarchy that requires not only

The twenty amino acids making up proteins.

Amino acid	Abbrevations	Amino acid	Abbreviations*
Glycine	Gly, G	Methionine	Met, M
Alanine	Ala, A	Tryptophan	Trp, W
Valine	Val, V	Tyrosine	Tyr, Y
Leucine	Leu, L	Asparagine	Asn, N
Isoleucine	Ile, I	Glutamine	Gln, Q
Phenylalanine	Phe, F	Aspartic acid	Asp, D
Proline	Pro, P	Glutamic acid	Glu, E
Serine	Ser, S	Lysine	Lys, K
Threonine	Thr, T	Arginine	Arg, R
Cysteine	Cys, C	Histidine	His, H

* The names of the amino acids are given in both three- and one-letter abbreviations.

Some proteins and their functions in organisms.

Protein	Function or occurrence
Keratin	Hair, finger- and toenails, feathers
Collagen	Skin, cartilage, tendons
Pigment of red blood cells (hemoglobin)	Oxygen transport
Insulin	Sugar metabolism
Antibodies	Immune response
Hormones and receptors	Response to the environment
Endorphins	Response to pain

Amino acid sequence of bovine insulin

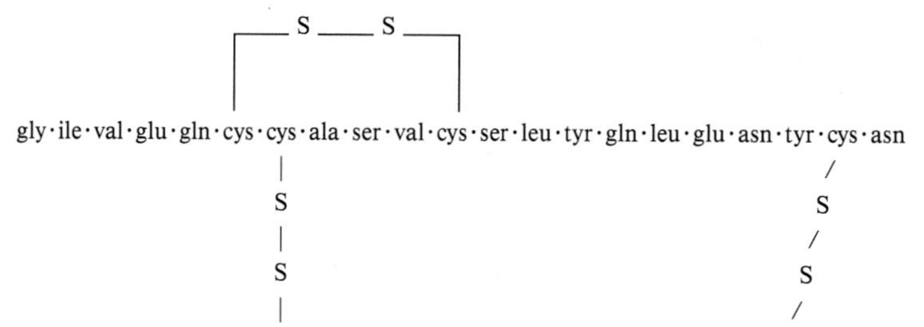

structural but also temporal dynamical coordination. In a healthy organism, cells divide at exactly the right moment. At a certain time during embryogenesis, the cells of the central nervous system distribute themselves between the head and the spinal cord in a precisely defined manner. The body stops growing taller at puberty. After a wound has healed, no further growth occurs. The follicle in the human female releases its ovum every four weeks. These are all highly complex, interconnected processes, which cannot be understood in static terms. They are part of a forward-directed, dynamical process. One hierarchy of order rises on top of another and the resulting dynamical system produces disorder. Wherever hierarchies of order are constantly being produced, there has to be a steady degradation and decay of structures, molecules, and fuels. But they do not decay spontaneously and willy-nilly. Their energy content is instead put to use to form new structures.

The decoding of a tape recording of music requires a specific decoding key and a decoding system. The oriented iron particles transfer their magnetic structure to the tape-recorder head, which then converts it electronically, via a modulated current of electricity, into the sound waves emitted from a pair of speakers. We hear these sound waves as music. This decoding process involves both translation and the consumption of large amounts of energy. To translate the information recorded on a

First base (5′ end) ↓	Second base U	Second base C	Second base A	Second base G	Third base (3′ end) ↓
U	Phe	Ser	Tyr	Cys	U
	Phe	Ser	Tyr	Cys	C
	Leu	Ser	STOP	STOP	A
	Leu	Ser	STOP	Trp	G
C	Leu	Pro	His	Arg	U
	Leu	Pro	His	Arg	C
	Leu	Pro	Gln	Arg	A
	Leu	Pro	Gln	Arg	G
A	Ile	Thr	Asn	Ser	U
	Ile	Thr	Asn	Ser	C
	Ile	Thr	Lys	Arg	A
	Met	Thr	Lys	Arg	G
G	Val	Ala	Asp	Gly	U
	Val	Ala	Asp	Gly	C
	Val	Ala	Glu	Gly	A
	Val	Ala	Glu	Gly	G

Fig. 1.18. The genetic code. Triplets of nucleotides (codons) in RNA are translated during protein synthesis according to the rules shown here. For example, the codon GUG is translated into valine and the codon GAG into glutamic acid.

magnetic tape – which, at least formally, has zero energy content – into the music produced by a 40-Watt loudspeaker, the tape recorder requires not just 40 Watts, transmitted ultimately as sound, but roughly ten times this quantity, or 400 Watts. Ninety percent of the energy "decays" to worthless, useless heat.

A similar process is involved in decoding the nucleic acids and translating their information into protein sequences. A complicated biochemical scheme, which will be described later, converts the information carried in the nucleic acids into protein sequences. This process is governed by the rules of the genetic code: three nucleic acid building blocks (nucleotides) code for each amino acid. The coding DNA thus contains three times as many building blocks as the corresponding proteins. This triplet code (Fig. 1.18) underlies the organizational scheme relating the informational content of DNA to the corresponding proteins. It was elucidated in 1962 by Marshall Nirenberg and Heinrich Matthaei.

Order in Living Systems Is Understandable in Principle. But Do We Really Understand It?

Some of the basic organizing principles of living systems have been presented in this introductory chapter. Life is capable of self-organization. The required instructions are stored as DNA and read off this tape of information. Is life a machine, then? Is it first wound up and then allowed to run automatically according to the intrinsic laws of the nucleic acid code? Did life arise by an act of creation?

Whatever the case may be, life does not merely involve action and reaction. It is a network, each part of which affects the whole. Furthermore, it is a dynamical network that changes in time and space. Under the same conditions and at the same point in space, different temporal events are possible. Likewise, under the same conditions and at the same point in time, different spatial events can take place. Goethe said to Eckermann: "It seems to hold true for systems that they are not completely accessible to themselves."

The maintenance of a high level of order in such dynamical systems is only possible through the constant formation of new structures. These compensate for the decay that is always associated with life. Only on its surface is the flow of life steady and unchanging. Beneath the macroscopic surface lie material and energy cycles without which life would cease to exist. Life is at once flowing and still.

Conrad Ferdinand Meyer

The Roman Fountain [10]

Up rises the jet and falling
Fills the marble bowl to its round brim,
Which, veiling itself, flows over
Into the depths of a second bowl;
The second, becoming too rich, gives,
Swelling up, its flood to the third,
And each takes and gives at the same time
And flows and rests

2. Biochemistry – Gain through Chaos

A dialogue between Georg Christoph Lichtenberg and Alice*, who inquire into what is real and what is false and come to the conclusion that nonsense can be informative, that chaos can be useful, and that one always gets somewhere if one only walks long enough[1].

ALICE: Professor, would you kindly explain to me what a "mock turtle" is? Recently, I met one that talked in such a curious way. I mean, Professor, would you tell me what a "mock turtle" really is?

LICHTENBERG: What do you mean by real, Alice? Can a mock turtle be real at all? *It often seems to me that "reality" is a purely mental construct. One thing I know for sure, though: If I wrote about this, the world would take me for a fool. Therefore, I prefer to remain silent. It is no more a subject to be discussed than the spots on my tablecloth are musical notes to be played on the violin. And yet many a scientist and artist has been inspired by spots on a tablecloth. Indeed, a cosmic number of such inspirations are possible.* Nonsense can be very meaningful. Obviously, a mock turtle is nonsense. There are turtles, there is turtle soup, there is even a mock turtle soup made from veal. But this does not imply that there is also a mock turtle. Think it over, Alice!

ALICE: But Professor, I met the Mock Turtle in Wonderland the other day. He was weeping so sadly and said, *with a deep sigh, 'I was a real Turtle.'* He told me, *'we went to school in the sea' and 'took the regular course': 'Reeling and Writhing, of course, to begin with, and then the different branches of Arithmetic – Ambition, Distraction, Uglification, and Derision.'*

LICHTENBERG: Good God! What a linguistic chaos! Your turtle *would surely choose the multiplication table as his patron saint!*

Granted that a mock turtle necessarily uses wrong expressions. This behavior is

---

* Alice is the main character in "Alice's Adventures in Wonderland" by Lewis Carroll (the pseudonym of C. L. Dodgson), who wrote the book for his seven-year-old friend Alice Liddell.

species-specific, as it were. Yet perhaps chaos is not without its underlying rules after all. Something reasonable, indeed something entirely new, could emerge from it. In fact, your sad mock turtle expresses more with his mumbo jumbo than with the correct words. Take alone uglification and derision. It is immediately obvious that he means multiplication and division. The information is all there. Yet this chaotic outburst conveys his emotional state of mind as well, this silly creature's ungrounded fear of our beautiful mathematics. It is truly astounding how chaos opens up a new level of communication and broadens the horizons of a system. *Perhaps even a slumbering dog or a drunken elephant has ideas worthy of a philosopher.*

In any case, your mock turtle suffers from some sort of psychological speech dysfunction. *It is possible to say that two times two is five. At least one can pronounce it. But how can one think it? For this very reason, I have often wished that there might be a language in which nothing wrong could be said. Or at least in which every violation of the truth would also be a grammatical mistake. But, of course, this would have its unfortunate side as well. Instead of the lively chat you two just had, the conversation would have been either very still or full of grammatical blunders. So it is better for you to talk nonsense.* Perhaps it is even a higher kind of nonsense, an *uncommon nonsense*, who knows?

ALICE: You are so intelligent, Professor. You make sense of even the most curious things. Have you ever done anything silly in your life? I mean, just for fun, as a game?

LICHTENBERG: Alas, my child, now you are getting me quite depressed. Just listen to this: *On the night of the 9th to the 10th of February, I dreamt that, while on a journey, I was dining at a country inn. Actually, it was a streetside tavern in which people were playing at dice. Opposite me sat a well-dressed, but somewhat dubious-looking young man, who was eating his soup, oblivious to the people standing and sitting around him. Every second or third spoonful of soup he tossed into the air, caught it again with the spoon, and then swallowed it calmly. What made this dream especially strange was my customary observation that such things could not be invented, but had to be seen. (That is, though no novelist could conceive of such a thing, I had imagined it at this very moment.) Near the dice throwers sat a tall, thin woman, knitting. I asked her what could be won in this game and she answered, "Nothing." And when I asked whether something could be lost, she said, "No!" Still, I considered this a very important game.*

ALICE: The man with the spoon is funny, but he seems a bit mad. Hmm, a game with nothing to win and nothing to lose. What good is that?

LICHTENBERG: The purpose of a game is not to win or to lose, Alice. I would consider you very egotistic if you were to think like this. The purpose of a game is simply the game itself: to introduce into our linear world the lively element of chance, the unpredictable dice, chaos. Without chaos, nothing new arises. Everything new and important is nonlinear.

ALICE: Nonlinear? Does that mean one cannot lay a ruler against it?

LICHTENBERG: Something like that. Neither a straightedge nor a French curve. There are twists and turns. Like thought, say, or fantasy. But also like a growing tree or a flower. Only idealized processes are linear. Reality is nonlinear; it springs about, it slips away, it deviates. Deviant, isn't it?

ALICE: Then, *uncommon nonsense* could be meaningful? My Cheshire Cat was exceedingly intelligent, even though it, too, was deviant, maybe even mad.

LICHTENBERG: Why was it mad, Alice?

ALICE: It told me so itself and even proved this logically beyond a doubt. It argued as follows: *"To begin with, a dog's not mad. You grant that? Well, then, you see a dog growls when it's angry, and wags its tail when it's pleased. Now I growl when I'm pleased, and wag my tail when I'm angry. Therefore, I'm mad."* I thought this was good logic and could offer no objections to it. What do you think, Professor?

LICHTENBERG: My dear child, you are really precocious. Though you are younger than my little Dörte*, she doesn't ask me such questions. *Perhaps it is even possible to make up a fairy tale based on Kant's philosophy. What do we know about the prerequisites for cognition in animals? There might even be an animal whose brain is the ocean and for whom the north wind is blue and the south wind red.*

ALICE: I have never met such an animal.

LICHTENBERG: Well, I think I should say farewell for now.

ALICE: *Would you tell me, please, which way I ought to go from here?*

LICHTENBERG: *That depends a good deal on where you want to get to.*

ALICE: *I don't much care where –*

LICHTENBERG: *Then it doesn't matter which way you go –*

ALICE: *– so long as I get somewhere.*

LICHTENBERG: *Oh, you're sure to do that, if you only walk long enough.*

Protein Biosynthesis – A Strategy to Avoid Chaos and to Gain Control

The task of DNA-directed protein biosynthesis is to select certain amino acids from among the twenty different ones available (cf. Fig. 1.18) and to link them together in a prespecified sequence. The resulting chain of amino acids, called a protein, should be identical to the billions of other proteins of the same kind. In other words, the amino acids have to be incorporated with as little error as possible. An error in

* Dorothea Stechard (nicknamed Dörte, 1765–1782), Lichtenberg's mistress since 1778.

only one protein molecule in a hundred, for example, would probably be tolerated by an organism. But even this level of accuracy demands precision work. To forge or synthesize a chain of one hundred amino acids with only one error in a hundred chains requires an accuracy of $1:100 \times 100 = 1:10000$.

Order in living systems is maintainable only if the building blocks of life, the amino acids, are selected accurately and assembled flawlessly to form proteins. Highly accurate recognition and selection of amino acids is clearly the key to order in life. But how are the chemical structures of amino acids recognized?

Nearly one hundred years ago, Emil Fischer published an article in *Chemische Berichte* with the title "Einfluß der Konfiguration auf die Wirkung der Enzyme" (Effect of Configuration on the Activity of Enzymes)[2].

Having carried out experiments on the specificity of enzymes, Emil Fischer wrote: "To use a metaphor, I would say that enzyme and substrate must fit together like lock and key in order to exert a chemical effect on each other. In any case, this notion becomes more likely and its value for stereochemical research increases when the phenomenon itself is transferred from the biological to the chemical realm."

These propositions make two assertions that were unorthodox at the time:
1. Biology at the molecular level becomes chemistry.
2. The substances of life fit together like lock and key.

Today there are countless confirmations of this theory. At one time we studied certain systems, the so-called inclusion compounds, that model very closely this concept of lock and key[3]. These compounds bind specifically and strongly to molecules that fit into their interiors. Chemical bonding is not required.

Two molecular locks of this kind are shown in Figure 2.1. The interior or cavity is formed by six glucose molecules and has an inner diameter of 6 Å. Accordingly, the cavity just barely accommodates a molecule of chlorobenzene; it is too small for larger molecules. This principle is the basis of the whole field of host-guest chemistry, where work has focused on modeling enzymatic catalysis[4]. There has been a recent resurgence of interest in this field[5].

But such molecular locks and keys are still not sufficiently accurate to ensure perfect order. In our analogy, skeleton keys and picklocks suffice to open these crude locks. These recognition systems are too simple to afford the accuracy of $1:10000$ or even greater required in the synthesis of proteins and nucleic acids.

We have examined the case of two very similar amino acids, isoleucine and valine[6], which differ by a single CH_2 group (Fig. 2.2). Valine, the wrong key, lacks just one ward and therefore fits into the isoleucine lock.

Theoretically, it can be shown that no normal enzyme, indeed no physical system in the world, is capable of discriminating between these two similar amino acids with an accuracy better than $1:5$[6]. This catastrophic situation would have rapidly led to chaos, to a collapse of protein interactions in living systems, had nature not developed a completely new principle of selection, a selection tree. The amino acid is checked for accuracy not only once but several times. The initial question of

Fig. 2.1. Inclusion compounds of α-cyclodextrin. Left: Empty cavity (lock). Right: Chlorobenzene inserted into the cavity.

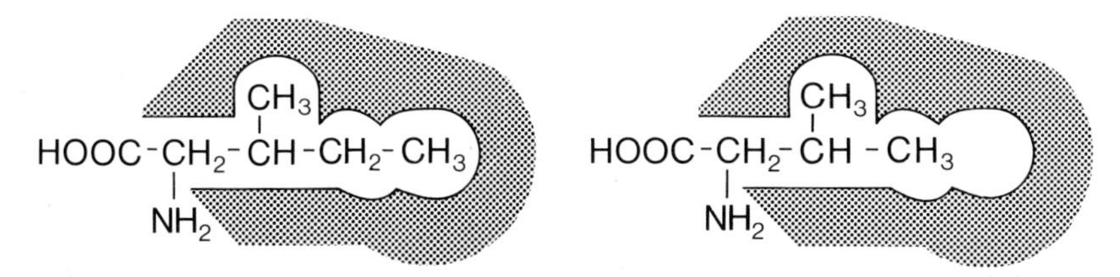

Fig. 2.2. Lock-and-key recognition of the amino acids isoleucine and valine in the "isoleucine lock." The "valine key" (right) can easily be inserted into the isoleucine lock.

the selection tree and those that follow are fundamentally different, however (Fig. 2.3).

The question is posed just once in the classical sense, that is, by inserting the key into the lock. If the substance is correct, it is processed further. If it is incorrect, it is rejected. Since the system is at thermal equilibrium, no energy is consumed and the process is reversible. This kind of process is operative in almost all enzymatic degradations and transformations describable by the so-called Michaelis-Menten theory. Protein biosynthesis, however, requires a higher level of accuracy than that allowed by simple physicochemical principles. Therefore, an additional process is employed and, if necessary, still further processes. After the amino acid is incorporated into the protein, a second question is posed: Was it correct? If yes, fine. But if the answer is no, the substance is discarded. In other words, a proofreading step has taken place, very similar to the proofreading of a printed text before it is published. This second process consumes additional energy; it is irreversible. Instead of a single selection step, then, a selection tree or cascade is involved: material flow – that is, chem-

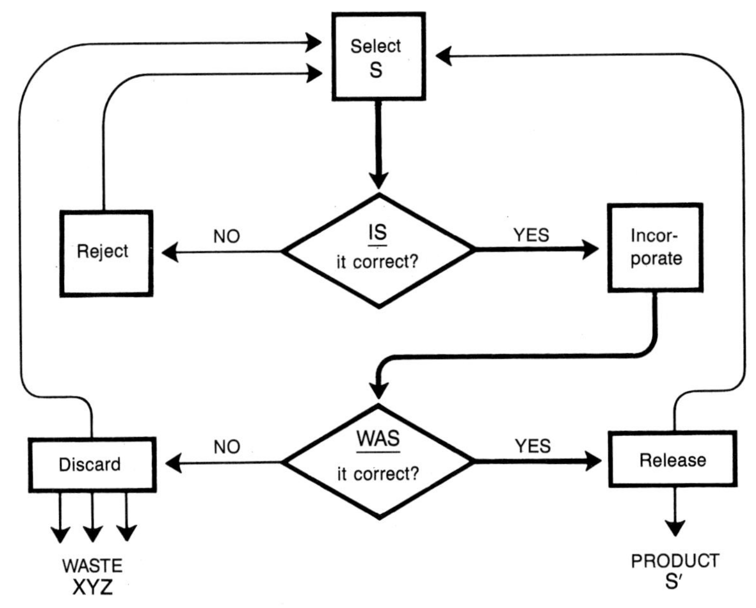

Fig. 2.3. Flow chart for the selection of amino acids used in protein biosynthesis. Whereas the first question is "reversible," the second is "irreversible" since energy is consumed.

ical transformation – and energy flow are directly coupled. Here, we encounter a totally new type of reaction, one that takes place far from equilibrium. Yet matter that is artificially kept in a state of nonequilibrium through expenditure of energy has entirely different properties from matter in equilibrium. Matter in equilibrium is uniform and uninteresting. Matter in a state of nonequilibrium is sensitive and highly specific. Indeed, living matter is created in just this way.

Amino acids are thus chosen via a selection tree that proceeds far from equilibrium (Fig. 2.4) at great material and energy expense. Only by bringing matter into a nonequilibrium state can isoleucine be discriminated from valine with an accuracy of one in roughly forty thousand. An accuracy of a mere one in five would be expected for a classical system[5].

This novel enzymatic system possesses the following features:

1. Branch points allow for feedback;
2. Energy is consumed to produce order;
3. The laws of classical thermodynamics have to be supplemented by those of nonequilibrium thermodynamics, since this selection process is dissipative or energy-consuming. Increasing the accuracy from 5 to 38000, for example, requires the expenditure of five additional molecules of ATP (adenosine triphosphate, the energy currency or fuel for every cell).

Protein biosynthesis thus employs the following strategy to avoid chaos: energy is consumed in order to transform chaos into order. More precisely, the consumption

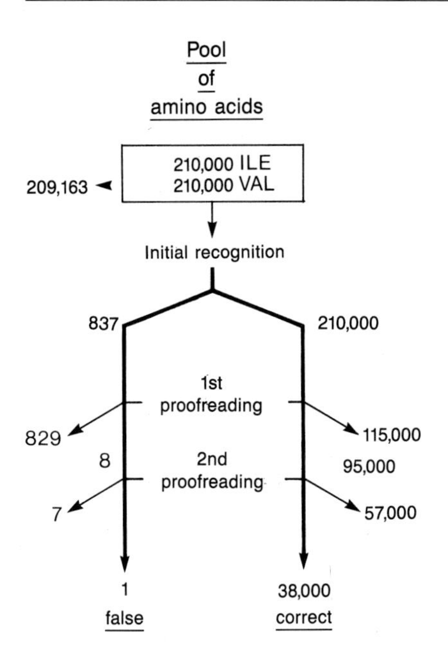

Fig. 2.4. Selection cascade for the amino acids isoleucine and valine.

of ATP improves the molecular recognition in this system by a factor of 10000. A selection tree involving dissipation of energy offers a further advantage; it is more easily regulated and fine-tuned. This principle is well-known in modern electronics. An electronic process can be fine-tuned by employing transistors, control elements wired in series, and feedback loops. Slight changes in the applied voltage are thereby sufficient to regulate even strong electrical currents. The biochemical process discussed here is similar. Slight changes in the reaction conditions lead to an increase in accuracy, accompanied by a corresponding increase in energy consumption. Alternatively, changing the reaction conditions in the opposite direction saves energy at the cost of a decrease in accuracy. This ability to modulate a biochemical reaction is potentially of great advantage to an organism. When nutrients are in short supply, it is able to synthesize new proteins with less accuracy, thereby conserving energy. For a short time, at least, this strategy should not result in undue damage to the organism. When nutrients are in full supply once again, it can resynthesize these proteins with high accuracy. The organism thereby becomes more adaptable and adjusts more readily to the conditions of its physical environment. This simple biochemical response reveals an important feature of living systems, their adaptability. These systems are adaptable because they are far from equilibrium. Moreover, the example clearly shows that the rigid, mechanical, Cartesian-Newtonian concept of matter is no longer adequate in the biological sciences. Matter far from equilibrium – and living matter is just that – has entirely new properties; it becomes adaptable, sensitive, even intelligent. This point will be discussed more fully in chapter 7.

Protein Quakes – Internal Strains Are Released in a Chaotic Fashion and Serve a Functional Role[7]

Myoglobin, like hemoglobin, is a protein that binds oxygen and makes it available for cellular respiration in muscles. This uptake and release of oxygen must be coordinated and regulated in a finely tuned manner to meet two requirements. On the one hand, the protein has to store and transfer oxygen; on the other, it has to release the oxygen immediately where respiration occurs. This is a problem of fine adjustment.

The protein consists of 173 amino acids, whose sequence is known. The three-dimensional structure of the molecule has also been elucidated. In fact, myoglobin was the first protein whose three-dimensional structure was determined by X-ray analysis in work carried out by J. C. Kendrew.

A heme group is embedded in the three-dimensional matrix of the molecule, just as it is in the blood pigment hemoglobin. Oxygen or carbon monoxide (CO) binds to the iron atom of the heme. To a first approximation, the molecule can be regarded as rigid; otherwise, an X-ray analysis of the three-dimensional structure would have been impossible to carry out. From a functional point of view, however, the molecule is not rigid. Binding and dissociation of oxygen or CO is accompanied by subtle structural changes, which are propagated throughout the entire molecule and are of central importance in fine-tuning the energy levels during oxygen binding.

Recent work has shown that these functionally important motions of the protein, and presumably of many other proteins as well, are nonequilibrium processes. They occur in the interior of the protein in analogy to earthquakes. Strain is released at the epicenter of the quake and propagated through the molecule as wavelike displacements and deformations. This is illustrated in Figure 2.5. Closer study of the energy states and wavelike displacements involved in this release of strain has revealed a fine-

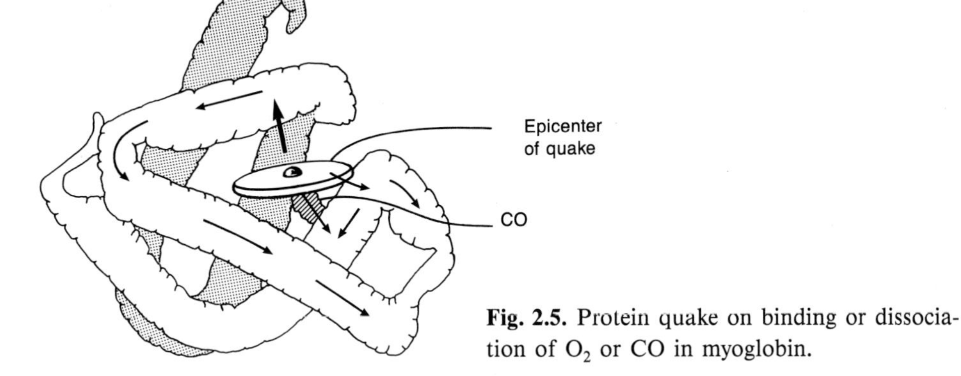

Fig. 2.5. Protein quake on binding or dissociation of O_2 or CO in myoglobin.

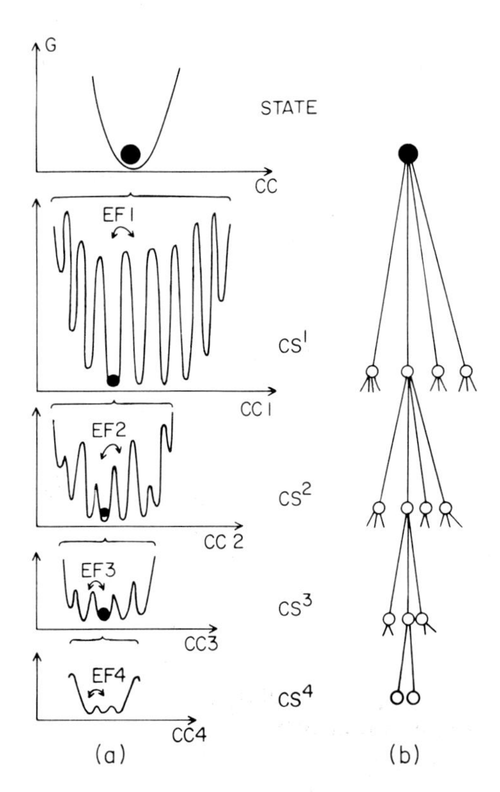

Fig. 2.6. Hierarchical arrangement of intermediate energy states on excitation of myoglobin: (a) schematic energy profile, (b) tree diagram of energy cascade.

ly tuned hierarchy of motions. The internal structure of the molecule does not directly attain the state of lowest energy, corresponding to the bottom of the deepest well. Once again, we are dealing with a finely tuned nonequilibrium system, like our hiker on the mountain ridge. In this case, however, there are a number of intermediate energy states (CS) that the molecule has to traverse and between which it oscillates. A short flash of laser light, sufficient to eject CO from the molecule, results in the molecular rearrangements shown in Figure 2.6. These rearrangements involve a cascade of coordinated motions. Accordingly, both binding and release of oxygen are finely tuned in a way that would have been impossible for an equilibrium process. The function of the molecule is explainable only in terms of this hierarchical protein quake.

Cells Talk to Each Other – But There Are Occasional Misunderstandings

An organism consists of many organs and various types of tissues, all of which have to be held together in a very specific way. Otherwise, there would be no difference between, say, connective tissue, cartilage, mucous membranes, glands, and so forth. The cells cannot simply be packed together like sandbags. They have to engage in some sort of specific contacts, as implied by the word tissue (from Latin *texere*, to weave). Yet little is known about these contacts. In this section I would like to speculate a bit and describe some of the results of our own research.

Fertilization – After Two Partners Have Found Each Other, Their Germ Cells Must Find Each Other, Too

During fertilization, the egg cell combines with a sperm cell. Each possesses a single set of chromosomes; that is, they are haploid. At the end of the fertilization process, the fertilized egg cell is diploid; it has two complete sets of chromosomes. How do spermatozoa find the egg cell? This biochemical process has been studied in lower organisms and in some higher organisms as well[8].

The spermatozoa first attach themselves to the egg cell. This is shown in Figure 2.7, where the egg cell of a mussel is surrounded by attached spermatozoa. This mutual encounter is a complex process. In the initial phase the recognition of a specific sugar structure on the surface of the egg cell plays a central role. The surface of every cell is covered with a thick layer of sugar molecules, some of them very complex. Here, we can apply the lock and key analogy once again. The sugar structures are the keys. As shown in the schematic drawing of a cell membrane in Figure 2.8, these sugar keys protrude from the membrane surface and serve as recognition elements. We now know that the head of the spermatozoon contains a lock called bindin, a molecule that mediates binding (Fig. 2.9). Bindin enables the head of the spermatozoon to attach itself to the egg cell. This lock and key combination is very specific. It corresponds to a very cleverly designed safety lock, which can be varied in a large, perhaps even infinite, number of ways. Fertilization is therefore species-specific. Following attachment through this sugar-protein contact, a series of events culminates in penetration of the egg by the head of a spermatozoon. The "lock" molecules mediating cell contact are also called lectins (Latin: *legere*, to read); they are able to read molecular handwriting. We have coined the name "cell-adhesion lectins" (CAL) for the types of molecules that appear to be the general mediators of highly specific cell-cell contact[9].

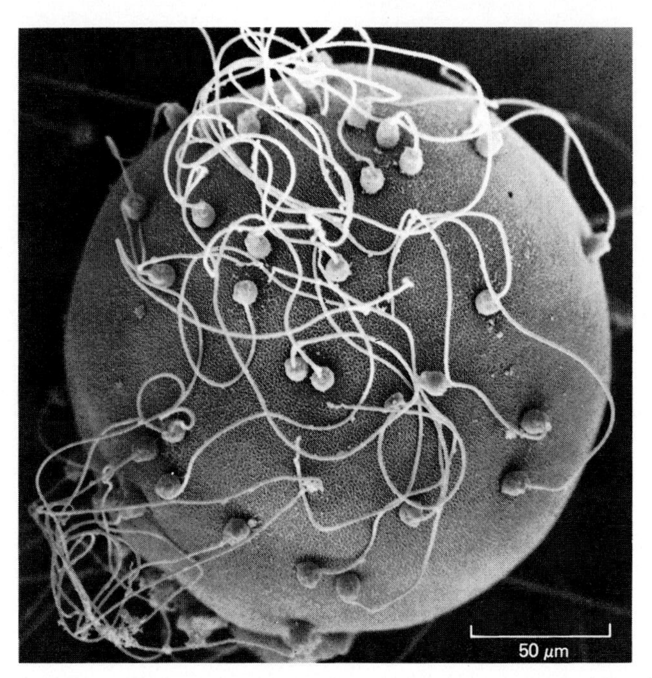

Fig. 2.7. Electron micrograph of a mussel egg with many spermatozoa attached to its surface.

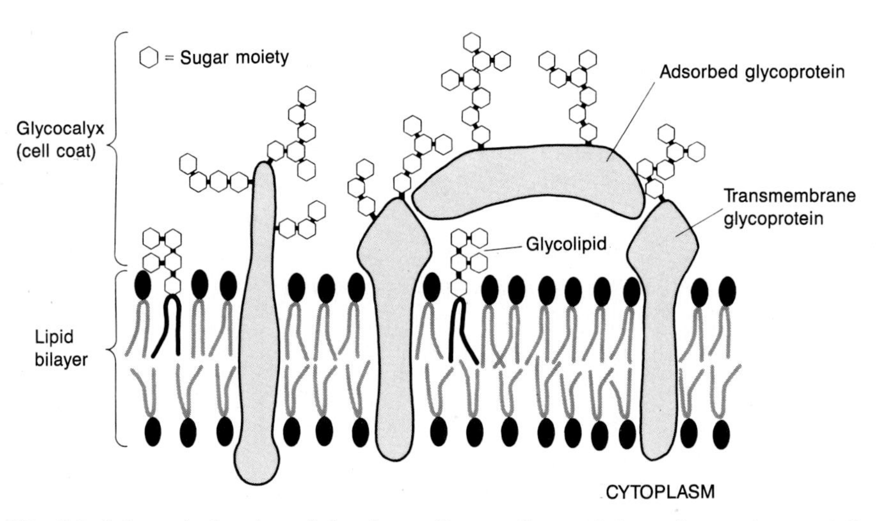

Fig. 2.8. Schematic drawing of the glycocalix or cell coat. It is made up of sugar (oliogosaccharide) side chains that are joined to intrinsic membrane glycoproteins and glycolipids as well as to glycoproteins and proteoglycans (not shown) adsorbed from outside. Note that all sugar moieties are located on the outer side of the membrane.

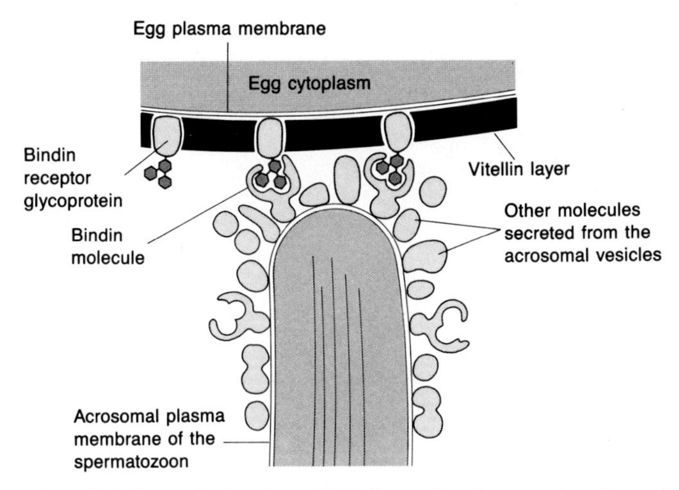

Fig. 2.9. Schematic drawing of bindin molecules covering the surface of the acrosome of a sea urchin spermatozoon. These proteins presumably bind to the vitellin layer of the egg cell through a specific oligosaccharide sequence.

Morphogenesis – Cell Surfaces Fit Together Like LEGO Blocks[10,11]

It has long been known that cell surfaces display certain sugar groups (glycoproteins) that are characteristic of each cell type. The best examples are the substances that determine blood groups. The three groups of red blood cells (A, B, and O) have different, mostly rare, sugars on their cell surfaces. The structure of these sugar moieties is under strict genetic control and hence cannot be unimportant. Nonetheless, it long remained a complete mystery why all cells possess these complex sugar structures (glycoproteins).

In 1985 we found that the cells of higher organisms have sugar-binding proteins on their membranes. These proteins, which exactly match their respective sugars, were previously known to be present only in plants, where they are called lectins. The occurrence of lectins in the cell membranes of higher cells has led to a completely new interpretation of glycoprotein patterns.

Glycoprotein patterns are tissue-specific. By examination of various tissues (liver, lung, pancreas, bone marrow), we showed that the corresponding lectins are also tissue-specific. We found, in addition, that cell-cell contact is mediated by glycoprotein-lectin interactions. These findings painted a whole new picture of the mechanism of interaction: Glycoproteins and lectins are like locks and keys (Fig. 2.10, top); cells are probably interconnected through many glycoprotein-lectin interactions. They snap together like LEGO blocks. This picture has been helpful in understanding tissue morphogenesis.

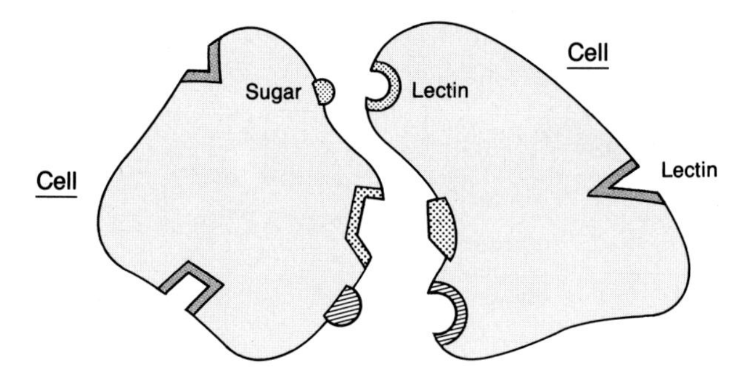

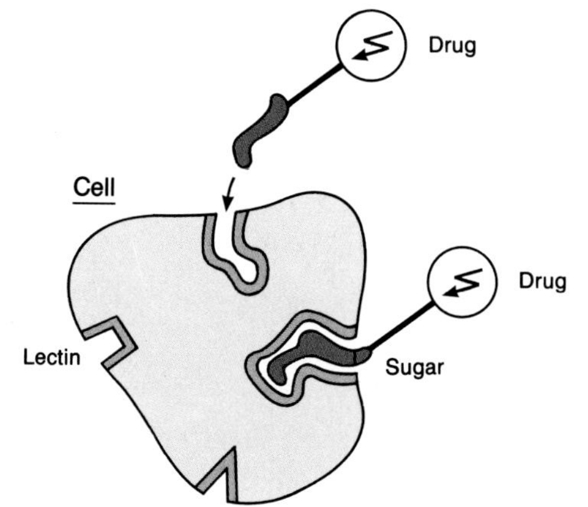

Fig. 2.10. Top: Schematic drawing of specific cell-cell contacts through sugar-lectin binding. Sugar residues (the keys or heads) protrude from the cell surface and fit firmly and specifically into the lectins (locks, holes). Below: A possible application of lectins in chemotherapy. A drug is coated with sugars matching the lectins of specific cells. In this way, the drug can be "addressed" to certain cells.

Cancer Metastases – The Invasion of Foreign Cells

Normally, tissue grows only where it belongs, precisely because of the specific cell-cell contacts through sugar-lectin binding. Cancer cells, however, have been shown to possess a characteristically modified pattern of lectins. The locks on their surfaces match the keys of other tissues[12,13]. To apply this finding to the study of tumors, the following possibilities or questions need to be considered:

1. *Are there tumor-specific lectins or lectin patterns?*

Each tissue, each cell group, appears to possess its own pattern of lectins. This leads to the question whether the different types of tumors also possess exactly defined patterns of lectins. From what is already known, the answer to this question is yes. The lectin pattern of a specific type of tumor is fundamentally different from that of the mother tissue in which this tumor first appeared and it is characteristic of the respective type of tumor.

2. *Can tumor-specific lectins be used for histological characterization?*
Tumor lectins, as sugar-binding proteins, may be exploited to stain tumor cells specifically. To accomplish this, a suitable dye has to be attached to a sugar that is specific for a particular lectin. Neoglycoproteins, proteins that are coupled to both a dye and a sugar, are used for this purpose. This staining agent is then employed to prepare a catalog of lectins with various sugar specificities. This catalog can be used, in turn, to characterize the different tumors of clinical importance. It is hoped that this work, carried out in collaboration with pathological institutes, will expand the array of diagnostic tools available to identify different types of tumors and to assess their degree of malignancy.

3. *Can tumor-specific lectins be employed in cytostatic therapy?*
The same principle used for staining could also be exploited to deliver cytostatics specifically to tumor cells (the magic bullet or Paul-Ehrlich principle). The principle is simple. An inert carrier is chemically prepared in such a way that it contains not only the specific sugar residue for the tumor lectin – its address, as it were – but also a cytostatic or other therapeutic agent. Upon delivery, the carrier undergoes degradation on the surface of or within the tumor cell and releases this agent. Since the carrier is taken up primarily by tumor cells, this approach, called drug targeting, could reduce the negative side effects of cytostatic chemotherapy (Fig. 2.10, bottom) and thereby lead to improved therapy.

4. *Is it possible to control metastasis?*
Preliminary results indicate that glycoprotein-lectin interactions could also be exploited to suppress the formation of metastases. If this process of "dropping anchor" indeed occurs through a sugar-lectin interaction, then it could be interdicted therapeutically by blocking the tumor lectins (locks) with neoglycoproteins (duplicate keys).

Nonlinearity of Cell-Cell Contacts – A Wisp of Chaos

Molecular surfaces like those shown in Figure 2.8 or Figure 2.10 display an extremely complex geometry. This is equally true for contact with the corresponding lectin. Since specific cell-cell contact probably requires not just one site of interaction but several, if not many, lock-key contacts (lectin-sugar interactions), the geometry of this process must be very complex indeed. Normally, a surface may be regarded as two-dimensional. But this is not so for irregular molecular structures, solid-state surfaces, or landscapes. Such structures have fractal (=*fract*ured/fraction*al*) dimensions. This subject will be discussed in detail in chapter 5. The deviation from integral dimensions is also a measure of the inherent chaos in a particular physical process. Contact between cells is probably 2.2-dimensional, perhaps even 2.5-dimensional; it thus involves strong elements of chaos. Without these, in fact, the precision of this dynamical docking would not be possible at all[14].

Mutation – Gain through Error

Because we observe life over only short periods of time, it seems constant to us. The same plants and animals surround us. Each spring the anemones bloom. Each June the June bugs fly. Each fall the apple trees drop their fruit. As we have seen, the hereditary factors of all living things are laid down in the DNA of the cell nucleus, in a lengthy tape of information consisting of 10^{10} base pairs in humans. During each cell division, all of these base pairs have to be copied accurately and passed along to the next cell generation. It is hard to imagine how this complicated, seemingly impossible task of precision work is accomplished.

In fact, neither nature nor the biochemist is capable of carrying out a copying process so exactly and so flawlessly. And yet even a single error could lead to total loss of activity of the gene in question. Accordingly, the system of information founded on nucleic acids and hereditary factors ought to collapse or should never have arisen in the first place. But it did arise and it is stable. What is the explanation of this fact?

This paradox leads initially to a different question. How constant is life in reality? We have discussed evolution several times already. It consists of a stepwise change in hereditary factors, each step involving a mutation, a change in a single base in the nucleic acid chain. The elementary event in evolution, then, is a mutation coupled with gene duplication. Without mutation there would be no evolution. A selection mechanism exploits this event, filtering out the suitable mutants and rejecting the unsuitable ones. Evolution is thus a process involving filtration of all possibilities made available by a large number of mutations. This filtration selects for a certain advantageous character or ability, one that improves the survival chances of the organism in a particular niche of nature.

Mutations – that is, functional changes in the nucleic acids – arise in many ways and occur constantly in all organisms. As already mentioned, an inherent cause of mutations is the limited accuracy of the copying process during cell division. In the course of copying, incorrect bases are inevitably incorporated – for example, adenine (A) instead of guanine (G) (cf. Fig. 1.16). As a result, all subsequent copies have the wrong sequence as well. There are also numerous external causes of mutation. High-energy radiation (gamma rays, X rays, cosmic radiation) zaps individual building blocks of the nucleic acids. UV radiation cross-links thymidine residues (T) in the nucleic acid strand. The building blocks are also modified by many chemicals.

For instance, nitrite and bisulfite, two simple chemicals, transform cytidine into thymidine. Alkylating agents result in the loss of guanine, in particular, from the chain. It is nothing short of miraculous, then, that life remains so constant in a world in which the mechanism for copying DNA is far from perfect, where we are bombarded on every beach by natural radiation – not necessarily from nuclear reactors – and where we are exposed to countless chemicals, even in our natural surroundings. It is not the occurrence of mutations that requires explanation; they arise readily enough. What needs to be explained is both the stability of the gene and the fact that the number of mutations is not even greater than what we actually observe.

The rate of mutation is held so low and life kept so constant because DNA is constantly being repaired. As we have seen already in chapter 1, the genetic information is stored in both strands of the nucleic acid. When an error, a mutation, occurs in one strand, a "repair crew" arrives at the site almost immediately and recognizes this abnormality. This repair crew is not a mere fiction. Repair enzymes constantly move along the DNA double strand, checking whether the double helix is in order. Every discrepancy in base pairing is revealed as an alteration in the geometry of the double strand. This is immediately recognized by the repair crew, which replaces the defective site. These repair enzymes are especially active during and immediately after DNA duplication (replication). But they also check the nucleic acid at other times both before and after replication.

How high is the actual error or mutation rate? It is strongly dependent on the organism, the length of its genome, and the rate of cell division. In the primitive bacteriophage Qß, which has a genome 3400 bases in length and possesses no repair mechanism during replication, an error occurs once in every four thousand bases. This accuracy presumably holds for simple replicases lacking repair mechanisms. For the human genome, however, it would lead to a catastrophe. In the human genome, only fifteen of the 10^{10} base pairs are exchanged per year. An unimaginable level of precision work, representing an accuracy of $10^9:1$! Producing, say, the drive shaft

Observed rates of exchange for the amino acid sequences of various proteins during evolution

Protein	Specific evolution time* in millions of years
Fibrinopeptide	1.2
Hemoglobin	6.1
Cytochrome *c*	21
Histone H4	600

* The specific evolution time is defined as the time required on average before one nonlethal amino acid exchange occurs in a protein 100 amino acids in length.

of a large turbine 10 meters in length with equal accuracy would require a precision of 1/100 μ (1 μ = 1/1000 mm), that is 1/100000 mm.

Modern sequencing methods have been applied to several key proteins to measure the rate at which amino acids have been exchanged over the course of evolution[15]. In the table on page 47, the so-called evolution times are given in millions of years. For example, the number 6.1 for hemoglobin means that it takes on average 6.1 million years of evolution before one of the 100 amino acids in hemoglobin is exchanged. This estimate takes into account only exchanges that are not "extinct," that is, changes in hemoglobin structure that were neutral or advantageous. Even so, this number indicates that the members of the primate family, which includes both humans and monkeys, possess almost identical hemoglobin. The histones, which are involved in packaging the DNA of higher organisms, have remained even more constant. They have a specific evolution time of 600 million years. This means that, because vertebrates have existed for only the last 500 million years, all vertebrates must have identical histones.

Living matter is made up of essentially the same building blocks, then, and the diversity of species is the consequence of just a few mutational changes in several key proteins. Accordingly, the living world is more stable than we might assume at first glance. In fact, its stability is virtually impossible to imagine. Yet the small number of deviations occurring through mutation, the random errors, are the driving force of evolution. We will later see that self-organizing, evolving systems are not unique to biology.

A crucial impetus for the study of evolving systems was a contest sponsored by the Swedish Royal Academy in 1889. The following question was posed: "How stable is our solar system?" This question refers to the stability of evolving mechanical systems. I call it the "Swedish Academy question" and will come back to it in chapter 6. A similarly fruitful and far-reaching question, like that posed for astronomy, may be formulated for biology as follows: How stable is our genetic system? Just as the former question ultimately led to an understanding of what gives rise to mechanical forms (chapter 6), the latter question, which I would like to call the "Göttingen Academy question" *, leads to an understanding of what gives rise to biochemical forms[16]. What appears at first glance to be an error – namely, the mutation resulting from a copying error or the chemical instability of the nucleic acid – is, in the final analysis, a gain in flexibility and adaptability. Indeed, it makes evolution of the genetic system possible. Gain through error.

* I call this the "Göttingen Academy question" because two members of the Göttingen Academy of Sciences (Manfred Eigen and Friedrich Cramer) have devoted their scientific work of the past ten years to answering this question.

Edmund Blunden

The Survival[17]

To-day's house makes to-morrow's road;
I knew these heaps of stone
When they were walls of grace and might,
The country's honour, art's delight
That over fountain'd silence show'd
Fame's final bastion.

Inheritance has found fresh work,
Disunion union breeds;
Beauty the strong, its difference lost,
Has matter fit for flood and frost.
Here's the true blood that will not shirk
Life's new-commanding needs.

With curious costly zeal, O man,
Raise orrery and ode;
How shines your tower, the only one
Of that especial site and stone!
And even the dream's confusion can
Sustain to-morrow's road.

3. Genes, Genetic Maps, Gene Therapy – A Problem of Complexity

A dialogue between Johann Wolfgang von Goethe and Charles Darwin on evolution and on what is natural and what is divine[1]

DARWIN: Most Honorable Herr Geheimrat, *you have proposed a law of compensation or equilibrium of growth, according to which nature is compelled, on the one hand, to be frugal and, on the other, to be generous,* as you say. This law serves as an important basis for my theory.

GOETHE: Thank you, reverend sir. Indeed, I employed the pairs of terms "*power and limitations,*" "*capriciousness and law,*" "*freedom and restraint,*" "*flexible order.*" Today you might call them "potential and conditional," "random variations and natural law," "chance and necessity," in short, "evolution."

DARWIN: Precisely, Herr Geheimrat, I believe that my theory uncovers the mechanism of evolution, which functions roughly as follows: *As many more individuals of each species are born than can possibly survive; and as, consequently, there is a frequently recurring Struggle for Existence, it follows that any being, if it vary however slightly in any manner profitable to itself, under the complex and sometimes varying conditions of life, will have a better chance of surviving, and thus be naturally selected. From the strong principle of inheritance, any selected variety will tend to propagate its new and modified form.*

As according to the theory of Natural Selection an interminable number of intermediate forms must have existed, linking together all the species in each group by gradations as fine as are our existing varieties, it may be asked, Why do we not see these linking forms all around us? Why are not all organic beings blended together in an inextricable chaos? With respect to existing forms, we should remember that we have no right to expect (excepting in rare cases) to discover directly connecting links between them, but only between each and some extinct and supplanted form.

GOETHE: As you know, reverend sir, I have been seeking such links and, if I may flatter myself, not without some success. In 1784, in the course of my extensive anatomical studies, I discovered the intermaxillary bone – a discovery which, regrettably, has been largely forgotten in your generation.

DARWIN (interjecting): Oh, not I, my dear Herr von Goethe. I mentioned this discovery explicitly several times in my main work.

GOETHE: That is good indeed. *I am well able to appreciate the encouragement I have received from abroad for some time now to take up the sciences once again. My beloved Germany is really quite strange in its ways; I have been keeping a diligent eye on the meetings announced and held by the Deutsche Naturforscher und Ärzte during the past three years to see whether something would affect me, move me, excite me, as a devoted observer of nature for the past fifty years. However, except for certain details, which really have only added to my knowledge, nothing has come of this. I have not been confronted by any new challenge. I have not been presented with anything new.*

But at my age I don't want to complain, and certainly not to you, reverend sir. Let me point out a difficulty in the acceptance of your theory. *It is natural for man to regard himself as the end of all creation and to view everything else only in relation to himself and only insofar as it serves him or is otherwise useful. He rules over the vegetable and animal world and – all the while devouring other creatures as fitting nourishment – he pays honor to his God and praises God's benevolence in looking after him so paternally. He takes milk from cows, honey from bees, wool from sheep. He assigns to things a useful purpose and at the same time believes they were created for that purpose. Indeed, it is beyond his imagination that even the tiniest herb might not have been placed there for him. If he has not yet identified its use, he believes that someday one will surely be found.*

And just as man thinks in general, so he thinks in particular, and he doesn't hesitate to carry over his customary view from life into science as well and to question the purpose and utility of even the individual parts of an organic being.

For a while this may work and he may make progress even in the sciences; but sooner or later he encounters phenomena where this narrow point of view is inadequate and where, without resort to higher support, he becomes embroiled in outright contradictions. These teachers of utility would probably assert that the ox has horns to defend itself. Yet I ask, then, why don't sheep have any? And when they do, why are they wrapped around their ears and thus of no use? It is something different, though, to say that the ox defends itself with its horns because it has them.

The question of purpose, the question "Why?," is by no means scientific. The question "How?" takes us a bit further, though. For when I ask, "How does the ox have horns?," this leads me to an examination of its organization and, at the same time, teaches me why lions have no horns and cannot have any.

For instance, man has two hollow spaces in his skull. The question "Why?" would not take us very far here, whereas by asking the question "How?," I learn that these empty spaces are vestiges derived from the animal skull; they are more strongly pronounced in such lesser organizations and have not yet been completely lost in man, despite his higher nature.

There is yet another barrier to understanding in the popularization of your theory: *the notion of origin is absolutely ignored. Thus, when we see something in the process of becoming, we are to assume that it already existed.*

DARWIN: How thoroughly you have considered my ideas, Herr Geheimrat! *As Natural Selection acts solely by accumulating slight, successive, favourable variations, it can produce no great or sudden modifications; it can act only by short and slow steps. Hence, the canon of "Natura non facit saltum," which every fresh addition to our knowledge tends to confirm, is on this theory intelligible. We can see why throughout nature the same general end is gained by an almost infinite diversity of means, for every peculiarity when once acquired is long inherited, and structures already modified in many different ways have to be adapted for the same general purpose. We can, in short, see why nature is prodigal in variety, though niggard in innovation. But why this should be a law of nature if each species has been independently created no man can explain.* All species find their place in a common tree of life, which is not only symbolic but also meaningful in historical terms as a unified system.

GOETHE: That is too mechanical for me, reverend sir. *There is no system in nature. It has – it is life and succession from an unknown center to unknowable limits. However, what it forbids in general, it allows all the more willingly in individual cases. Each unique natural creature describes – in addition to the great cycle of all life in which it takes part – a narrower, more characteristic path. And the features of the latter which, all deviations aside, express themselves in the ongoing sequence of individuals in one cyclic path as well as another – this persistent recurrence in the course of changing appearances – marks the species. I am firmly convinced that what is of the same lineage is of the same species. It is impossible for one species to arise from another, for nothing interrupts the successive connections in nature; only what originally was placed side by side exists separately.*

DARWIN (to himself): Oh, my God, what effusive Teutonic gibberish! (aloud) Most Honorable Herr von Goethe, *what limit can be put to this power, acting during long ages and rigidly scrutinising the whole constitution, structure, and habits of each creature – favouring the good and rejecting the bad? I can see no limit to this power, in slowly and beautifully adapting each form to the most complex relations of life. The theory of Natural Selection, even if we look no farther than this, seems to be in the highest degree probable.*

GOETHE (coldly): *To try to explain the simple in terms of the compound, what is easy by what is difficult, is a grave illness pervading the body of science and, though perhaps recognized by those with insight, it is not always acknowledged.*

DARWIN: The mechanism of my theory, which we still do not know in detail, may be complicated. But the theory itself is quite simple, Herr Geheimrat. Naturally, the tree of life possesses branch points, so-called bifurcations, where developments diverge. Only in this way can something new arise. *A grain in the balance may determine which individuals shall live and which shall die – which variety or species shall increase in number, and which shall decrease, or finally become extinct. As the individuals of the same species come in all respects into the closest competition with each other, the struggle will generally be most severe between them; it will be almost equally severe between the varieties of the same species, and next in severity between the species of the same genus. On the other hand the struggle will often be severe between beings remote in the scale of nature. The slightest advantage in certain individuals, at any age or during any season, over those with which they come into competition, or better adaptation in however slight a degree to the surrounding physical conditions, will, in the long run, turn the balance.* Evolution is not simply an equilibrium system.

How many generations it takes for such a transition to occur between species we cannot yet say. But with your spectacular discovery of the intermaxillary bone you yourself have mapped the way (Goethe, reconciled, nods his head in agreement).

Organs in a rudimentary condition plainly show that an early progenitor has the organ in a fully developed condition; and this in some cases implies an enormous amount of modification in the descendants. Throughout whole classes various structures are formed on the same pattern, and at a very early age the embryos closely resemble each other. Therefore I cannot doubt that the theory of descent with modification embraces all the members of the same great class or kingdom. I believe that animals are descended from at most only four or five progenitors, and plants from an equal or lesser number.

Analogy would lead me one step farther, namely, to the belief that all animals and plants are descended from some one prototype. But analogy may be a deceitful guide. Nevertheless all living things have much in common, in their chemical composition, their cellular structure, their laws of growth, and their liability to injurious influences. We see this even in so trifling a fact as that the same poison often similarly affects plants and animals; or that the poison secreted by the gall-fly produces monstrous growths on the wild rose or oak-tree. One is compelled to conclude from all this that the basic biochemical processes have been the same since the origin of life. You yourself once wrote about the "primordial plant." This work, too, I have much admired, my dear Herr von Goethe.

GOETHE (now in a jovial mood, ruffling through old letters): Indeed, in 1787 I wrote from Italy to my mistress in Weimar: "*...that I am very close to the secret of the generation and organization of plants...The primordial plant will be the most wondrous creation in the world. Nature herself will envy me. With this model and the key it will then be possible to invent plants endlessly...the same law will be applicable to all other living things.*"

When facing primordial phenomena, when they appear unveiled to our senses, we are overcome by a kind of timidity, even fear. The sensuous person resorts to wonder.

When a body of knowledge is mature enough to become a science, a crisis necessarily results; for an apparent difference emerges between those who discriminate and describe each individual separately and those who keep an eye on the general and prefer to add or incorporate the unique. This is probably what happens to many, including myself, the first time they confronted your theory of evolution. Please pardon my slight ill-feeling a few minutes ago, reverend sir!

DARWIN (animated, almost beyond himself): Never mind, honorable sir, that is just the nature of scientific discussion.

It can hardly be supposed that a false theory would explain, in so satisfactory a manner as does the theory of Natural Selection, the several large classes of facts above specified. It has recently been objected that this is an unsafe method of arguing; but it is a method used in judging of the common events of life, and has often been used by the greatest natural philosophers. The undulatory theory of light has thus been arrived at by Newton.

GOETHE (taking offense, responding vehemently): I beg your pardon, reverend sir, you are a biologist; *please leave Newton out of this. You need know absolutely nothing about his Optiks. It is quite stupid and you wouldn't believe what damage is inflicted on an intelligent mind that considers such nonsense. Don't pay any attention to the Newtonists. Just be satisfied with my pure theory and you will be well-off.*

DARWIN (embarrassed, realizing his serious mistake): Pardon me, Most Honorable Geheimrat, it just slipped out. I am well aware of your theory of color and its great merits. I did not mean to venture into the field of physics. If you will listen to me for just one more minute, I will try to counter some of the arguments forwarded against my theory. *It is no valid objection that science as yet throws no light on the far higher problem of the essence or origin of life. Who can explain what is the essence of the attraction of gravity? No one now objects to following out the results consequent on this unknown element of attraction; notwithstanding that Leibnitz formerly accused Newton of introducing "occult qualities and miracles into philosophy."*

I see no good reason what the views given in this volume should shock the religious feelings of any one. It is satisfactory, as showing how transient such impressions are, to remember that the greatest discovery ever made by man, namely, the law of the attraction of gravity, was also attacked by Leibnitz, "as subversive of natural, and inferentially of revealed, religion." A celebrated author and divine has written to me that "he has gradually learnt to see that it is just as noble a conception of the Deity to believe that He created a few original forms capable of self-development into other and needful forms, as to believe that He required a fresh act of creation to supply the voids caused by the action of His laws."

GOETHE: You are absolutely right, reverend sir. *One must cleanly distinguish what is natural science, what is human, and what is divine. Yet these three areas of our being act upon and interact with one another in a splendid way. Without my undertakings in the natural sciences, I would never have come to know mankind as it is. No other thing can so closely approach pure observation and reasoning, the fallacies of the senses and the mind, the strengths and weaknesses of character; everything else is more or less flexible and vacillating, more or less open to consider-*

ation. But nature doesn't play games; it is always true, always serious, always strict, always right. The errors and fallacies are always human. Nature spurns the inaccessible. It only surrenders and reveals its secrets to what is accessible, true, and pure.

Reason is insufficient to grasp it. Man must be capable of attaining the highest level of reason in order to come even close to the divinity, which reveals itself in elemental phenomena, both physical and moral, behind which it is present and which emanate from it.

(Darwin grows restless, but restrains himself.)

The divinity acts in the living, however, not in the dead; it is present in what is becoming and what is changing, not in what has already become and is rigid.

Mister Darwin, how do you think your theory will develop further? What are its prospects?

DARWIN: *In the future I see open fields for far more important researches. Psychology will be securely based on the foundation already well laid by Mr Herbert Spencer, that of the necessary acquirement of each mental power and capacity by gradation. Much light will be thrown on the origin of man and his history.*

The other and more general departments of Natural History will rise greatly in interest. The terms used by naturalists, of affinity, relationship, community of type, paternity, morphology, adaptive characters, rudimentary and aborted organs, etc., will cease to be metaphorical, and will have a plain signification. When we no longer look at an organic being as a savage looks at a ship, as something wholly beyond his comprehension; when we regard every production of nature as one which has had a long history; when we contemplate every complex structure and instinct as the summing up of many contrivances, each useful to the possessor, in the same way as any great mechanical invention is the summing up of the labour, the experience, the reason, and even the blunders of numerous workmen; when we thus view each organic being, how far more interesting – I speak from experience – does the study of Natural History become!

A grand and almost untrodden field of inquiry will be opened, on the causes and laws of variation, on correlation, on the effects of use and disuse, on the direct action of external conditions, and so forth. The study of domestic productions will rise immensely in value. A new variety raised by man will be a more important and interesting subject for study than one more species added to the infinitude of already recorded species. Our classifications will come to be, as far as they can be so made, genealogies; and will then truly give what may be called the plan of creation.

GOETHE: My question about prospects really implied the practical side, the daily utility for mankind. *It is absolutely impossible to predict what the coming years will bring us; but I fear we will not achieve peace in the near future. It is not in the nature of the world to be content with itself. The powerful are unlikely to ensure that power is not misused. The masses are unlikely to be content with gradual improvements in their humble lot. If mankind could be made perfect, then a perfect state would be conceivable. As things stand, there is bound to be an eternal fluctuation to and fro.*

One part suffers while the other part prospers. Egoism and envy will always play the devil.

DARWIN (aside): Typical German "Weltschmerz"! (then aloud) It is beyond my abilities to estimate the practical utility that my theory of evolution will have for the cultivation of plants and the breeding of animals. As you know, I cultivate roses. It is just a personal hobby, though. My interest in the origin of new species, on the other hand, is purely scientific. *There is grandeur in this view of life, with its several powers, having been originally breathed by the Creator into a few forms or into one; and that, whilst this planet has gone cycling on according to the fixed law of gravity, from so simple a beginning endless forms most beautiful and most wonderful have been, and are being evolved.*

GOETHE: Only the seed of life from the creator? That is again too mechanical for me, reverend sir. *And anyway, what is all this and what does it mean? After the six imaginary days of creation, God by no means rested; rather, he has remained just as active as at the beginning. This crude world composed of simple elements and turning in the rays of the sun, year in year out, would certainly have given him less pleasure if he had not planned to establish on this material foundation a botanical school for a world of minds. Thus, he is now constantly active in higher natures in order to cultivate the lesser ones.*

Now, I am an old man almost 83 years of age*. My dear Mister Darwin, you are exactly 60 years younger than I. It is no small wonder that we assess the importance of science differently.

Farewell! (Darwin bows and leaves.)

Genes Can Be Taken Apart and Put Back Together

Genes are the smallest material units of inheritance, the atoms of inheritance. This does not mean, however, that we cannot take genes apart still further. They consist of a sequence of nucleic acid building blocks that can be disassembled with the help of chemical reagents. The "atomic" property thus pertains only to the biological phenomenon of heredity, not to the underlying chemical structures. Moreover, the individual building blocks are themselves composed of carbon, hydrogen, nitrogen, oxygen, and phosphorus atoms. Here, the atomic theory of the elements holds, although, as we know, even the atomic nucleus is made up of building blocks.

---

* The sentences in the preceding paragraph were written down by Eckermann eleven days before Goethe's death.

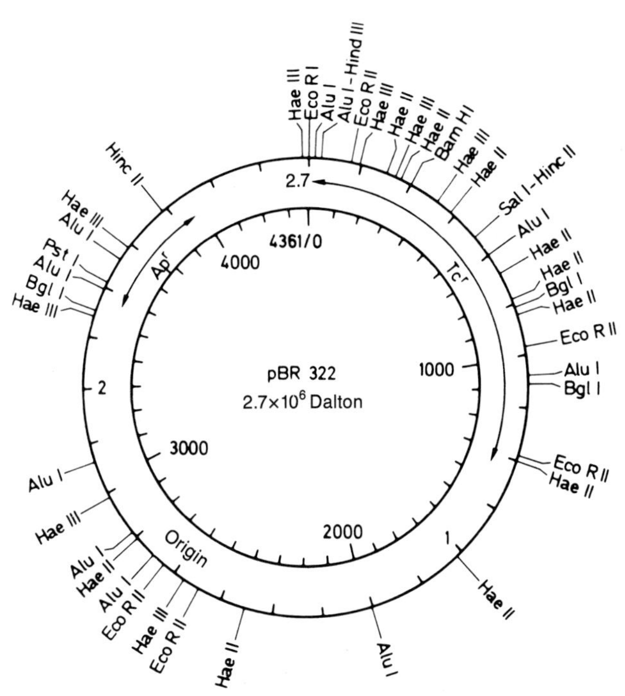

Fig. 3.1. The restriction genetic map of pBR 322, the best-known plasmid. B and R are the initials of the two people who constructed this plasmid, Bolivar and Rodriguez. The sites of cleavage by restriction enzymes are shown around the outer circle. The inner circle gives the number of base pairs (bp) starting from the Eco R1 cleavage site. The sites of the ampicillin and tetracyclin resistance genes (Ap^r and Tc^r) are shown between the two circles, together with the starting point or origin of replication. The cleavage sites are numbered with Roman numerals.

It is quite easy to disassemble a nucleic acid into its individual building blocks by a chemical process known as hydrolysis, that is, cleavage by water. Purely chemical hydrolysis, though, severs the long molecular chain rather arbitrarily. In order to determine the structure of the nucleic acid, it is necessary to cut the molecular thread at a relatively small number of very specific sites. So-called restriction enzymes are used for this purpose[2]. These enzymes are specific for certain sequences along the nucleic acid chain and cut at only the sites displaying these sequences. A nucleic acid with numerous sites of possible cleavage is shown in Figure 3.1.

This circular nucleic acid, called a plasmid, is used to incorporate genes into bacteria. The numerous sites indicated around the outer ring can be cut in such a way that a foreign gene may be inserted between the open ends. The choice of restriction enzyme determines the site at which this cutting occurs. These pieces of nucleic acid, which replicate independently in bacteria and higher organisms, are called plasmids

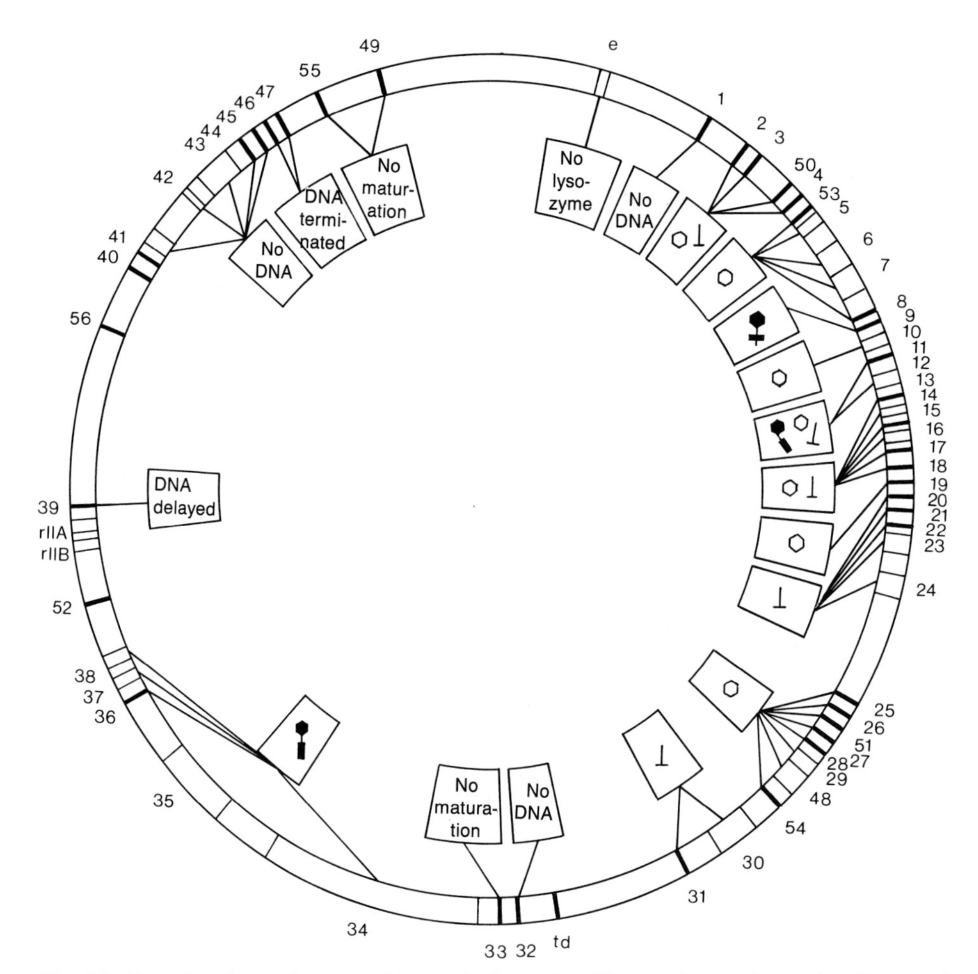

Fig. 3.2. Functional genetic map of bacteriophage T4. The numbers refer to the 56 genes in which conditional lethal mutations have been localized. The functions of the respective genes are given in the boxes; for example, no DNA means that this mutation prevents DNA synthesis from taking place. The boxes thus indicate what changes occur in bacteriophage T4 when the nucleic acid at a particular site is modified[3].

or vectors (from Latin *vehere*: to carry) because they serve as vehicles or vectors for other genes. Specific cleavage of these and many other bacterial and higher genes allows them to be localized on the nucleic acid strand; a genetic map is thereby obtained. The genetic map of a very simple organism, bacteriophage T4, is shown in Figure 3.2.

Plasmids may also be employed to smuggle foreign genes into cells, where they are then replicated. Under suitable conditions these genes are even incorporated into the actual genome of the host organism. Genes and groups of genes can thus be taken apart and put back together. The techniques for this are at hand for any chemistry or biology student to learn.

Genes Can Be Read

The nucleic acid sequence can be determined by a similar approach involving chemical degradation methods (sequential chain degradation or specific chain cleavage) and radioactive labeling[4,5]. To do this, the chain is cut once after each nucleic acid building block and the chain end is specifically labeled. Application of an electric field then orders the resulting chains according to length. With a little practice the sequence can be read off directly from the resulting "fingerprint" (Fig. 3.3[6]). Of course, this is easier said in a single sentence than done in the laboratory. In fact, it requires both theoretical knowledge and a certain level of skill, acquired only through practice, just as a watchmaker or a diamond cutter must learn their trade. To determine the sequence shown in Figure 3.4, four people worked for about one year at the Max Planck Institute for Experimental Medicine in Göttingen.

Longer genes are first cleaved to give restriction fragments, which are then sequenced. The resulting sequences constitute a puzzle, which must be fit together to

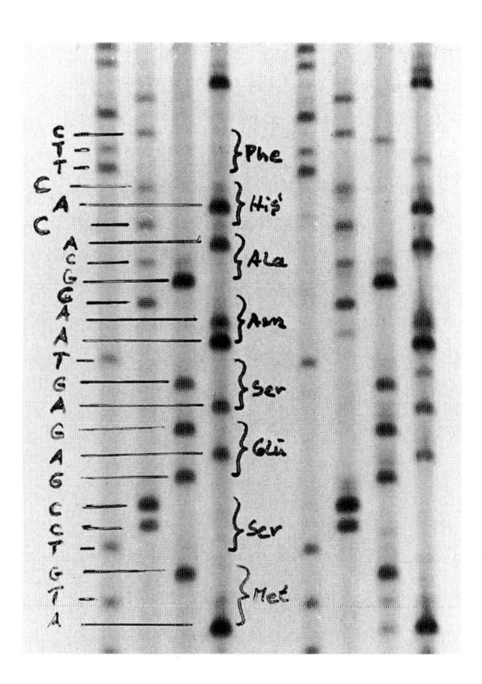

Fig. 3.3. Sequencing gel of a fragment of the gene for isoleucyl tRNA synthetase. The sequencing was performed according to the method of F. Sanger[5]. This fragment is the beginning of the coding part, namely, nucleotides 1 to 24, corresponding to amino acids 1 to 8 (cf. Fig. 3.4)[6].

```
-300 CCCAACTCGGGTGCTACGGGAGGTGGAGAAGATACAGGTCCAACAGTGGAAGAGTTGATTGATTATTCTTCTATA
-225 AGTGTTCTATTTAGCTACTTTTTATGTTTAACCTTTTATACGATGGCGGGTAATCTATCCATGATGACGAAAAAT
-150 TTTTTTTTTTTTTGTTTCCGCAGCACGCAAGAAATCTCGAAACAATGATGACTCTTAAGCATGAAAAATATCATT
 -75 TTGCGCTTTAAACTAGATTGATGTTACTCGACTTCCTACAACCTTTAGCCAAAAGCTTCAAAAAACCAAGGAAAT

  +1 ATGTCCGAGAGTAACGCACACTTCTCATTTCCAAAGGAGGAAGAAAAAGTTCTATCTCTTTGGGATGAAATAGAT
   1 MetSerGluSerAsnAlaHisPheSerPheProLysGluGluGluLysValLeuSerLeuTrpAspGluIleAsp

 +76 GCCTTTCATACTTCATTAGAATTAACAAAAGACAAACCGGAGTTTTCCTTCTTCGATGGGCCTCCATTTGCCACC
  26 AlaPheHisThrSerLeuGluLeuThrLysAspLysProGluPheSerPhePheAspGlyProProPheAlaThr
-.-.-.-.-.-.-.-.-.-.-.-.-.-.-.-.-.-.-.-.-.-.-.-.-.-.-.-.-.-.-.-.-.-.-.-.-.-.-.-.-

+3001 AAGAAGTGTGGTTTGGAAGCCACCGACGATGTTTTAGTGGAGTACGAATTAGTTAAAGATACTATCGACTTTGAA
 1001 LysLysCysGlyLeuGluAlaThrAspAspValLeuValGluTyrGluLeuValLysAspThrIleAspPheGlu

+3076 GCCATTGTCAAAGAACATTTTGATATGTTAAGCAAGACCTGTAGATCCGACATTGCCAAATATGACGGCTCAAAG
 1026 AlaIleValLysGluHisPheAspMetLeuSerLysThrCysArgSerAspIleAlaLysTyrAspGlySerLys

+3151 ACAGACCCAATTGGTGATGAAGAACAATCTATTAATGACACCATTTTCAAATTAAAAGTGTTCAAATTATGAAAA
 1051 ThrAspProIleGlyAspGluGluGlnSerIleAsnAspThrIlePheLysLeuLysValPheLysLeu***

+3226 CAACTCATATAAATACGTACAAATTTTTCTCTACTCGAAGTGATATAGATGTATATGTGTAAGTTTACGTTTAAG
+3301 ATTAGAGTCATGTAATGCTAACTGTCTCCACCGATAATGTTGTATAATACCCGTGAAATCATAGCACATGATATA
+3376 TCATCACCCGGAGGCCGGTTATTTTCGGCGGCGGCAAAAATATTTGGTATAATTATGGAAATACAAAAAGGGGAA
+3451 CCATTAAAGGTTGAGGAGGGGATTGATAAGAGAATCTAATAATTGTAAAGTTGAGAAAATCATAATAAAAATAAT
+3526 TACTAGAGACATGAAGTCTAC
```

Fig. 3.4. Nucleic acid and amino acid sequence of isoleucyl-tRNA synthetase. The one-letter abbreviations for the first 450 bases are shown at the top and those for the last 526 bases at the bottom. There are a total of 3826 bases in the sequence, the coding part of which extends from nucleotide 1 to 3222.

reconstruct the complete sequence of the nucleic acid. The longest piece of nucleic acid sequenced so far is that of bacteriophage lambda with 46000 base pairs. The method has been worked out with such precision that, when two laboratories determined the sequence independently, they found only one discrepancy, which was later accounted for.

Part of the sequence for isoleucyl-tRNA synthetase is shown in Figure 3.4. This enzyme is responsible for incorporating isoleucine into all proteins and thus performs the proofreading step discussed on page 35 in chapter 2[7]. The entire sequence of 3826 bases was determined and should be read left to right, row by row, just as a normal text is read. Immediately below the DNA sequence is the corresponding protein sequence derived unambiguously from the genetic code. The actual coding sequence begins after the first 300 bases. Written in this way, the information carried in the entire human genome would fill 500000 pages of text or one-thousand 500-page volumes. A worldwide effort is now underway to decipher it.

Genes Can Be Synthesized in a Test Tube

The basic chemical structure of nucleic acids is known and the sequence of specific nucleic acids can be "read." The foundation has thus been laid to construct a macromolecule from the individual building blocks by purely chemical means, that is, without the use of microorganisms or enzymes. This task is extremely complex and increases exponentially in difficulty with the chain length of the gene being synthesized. But the problem is solvable in principle, particularly since, as described already, fragments can be cut out of a gene and reinserted into it. It is not necessary to synthesize the entire gene, but only the part one wishes to change. A battery of procedures have been developed in recent years for the chemical synthesis of gene building blocks that can be inserted into mutant genes as probes or used to introduce new properties[8]. These procedures are in part so well automated that a programmable machine is capable of synthesizing virtually any gene fragment overnight. Chemical methods alone suffice to synthesize short oligonucleotides – that is, nucleic acid chains – up to about 150 bases in length. Enzymatic coupling of these chains then affords nucleic acids of virtually any chain length. Usually, though, it is not necessary to prepare long nucleic acids in order to modify genes, since a single modified building block already constitutes a mutation. If it is smuggled into the cell and undergoes replication there, the modified genetic characteristic (mutation) is passed on. Of course, insertion of longer pieces of nucleic acid is also possible. There is no theoretical limit, in fact, to joining various fragments of nucleic acid, be they natural or synthetic. But the problem can become quite complex in practice.

Genes Can Be Modified and Put Back into an Organism

What has been said so far shows that genes can be specifically excised, modified, and reinserted. Indeed, if both the sequence and function of a gene are known in sufficient detail to predict which functions will be affected by a specific mutation, then it should be technically feasible to prepare genes coding for proteins previously unknown to an organism. In this way, human insulin and somatostatin, the human growth hormone, were prepared in *Escherichia coli* (see table on page 63). A schematic depiction of site-directed mutation is shown in Figure 3.5.

Genes can thus be modified at will. The gene products, the proteins, are often so complex in structure and function, however, that it may be impossible to predict the ultimate effect of a given modification.

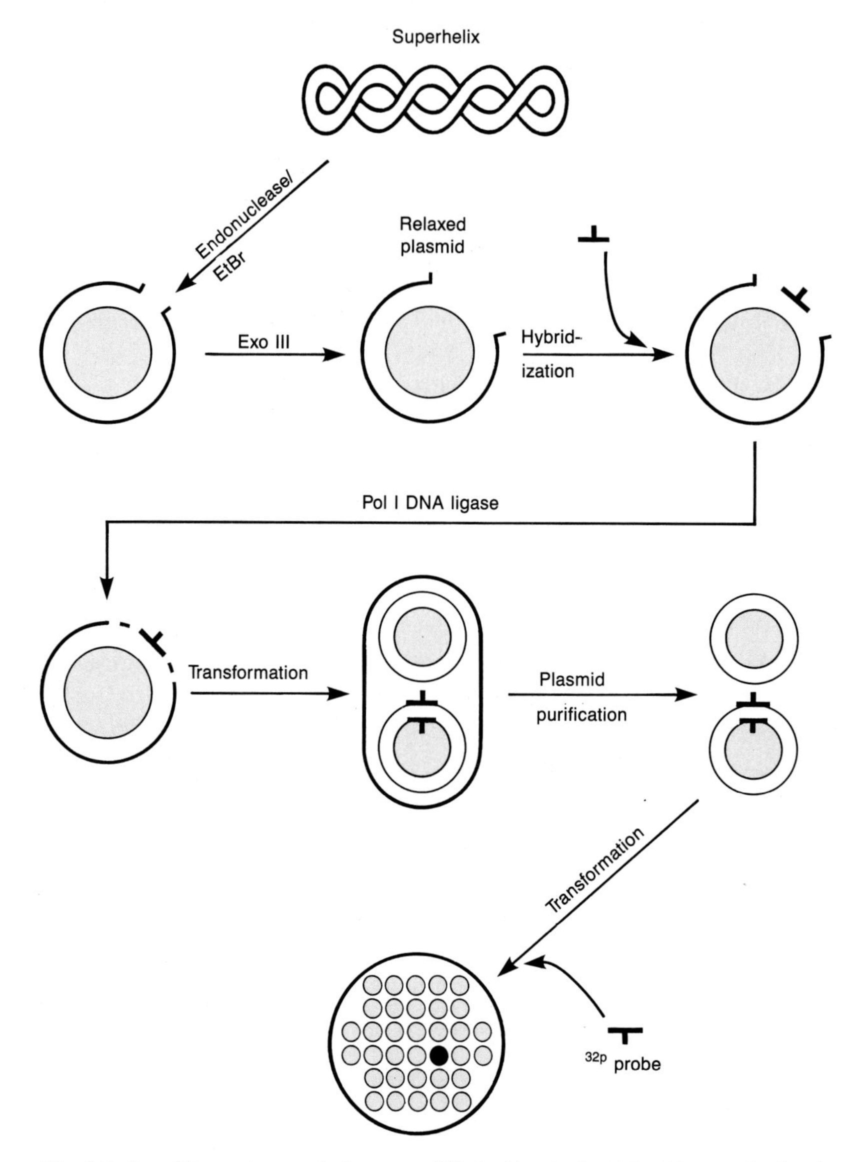

Fig. 3.5. Specific mutagenesis by a modified oligonucleotide. The synthetic piece of nucleic acid is inserted into the plasmid through hybridization. The plasmid then undergoes biological replication.

Some human proteins produced in bacteria.

Protein	Function
Insulin	Regulates sugar metabolism
Growth hormone	Promotes growth
Interferon	Antiviral agent
Parathyroid hormone	Regulates calcium metabolism
Somatostatin	Regulates production of growth hormone
Relaxin	Relaxes the uterus
β-Endorphin	Relieves pain
Factor VIII	Regulates blood clotting
Urokinase	Dissolves blood clots
Superoxide dismutase	Inhibits inflammation
Hepatitis B antigen	Vaccine against hepatitis
Tumor necrosis factor	Inhibits the growth of tumor cells

Genetic Engineering – What Is Technically Feasible? – A Positive and a Negative Catalog

The following prospects of genetic engineering or recombinant DNA technology are discussed in the context of foreseeable or feasible extensions of what has been accomplished already. In many cases, developments twenty or thirty years down the road are already in sight despite the breathtaking pace of current work. Of course, some things will probably never become feasible.

Modified Functions – New Enzymes

We have seen above that enzymes, the gene products, can be modified and tailored to serve new functions. A very simple example is the construction of protein-degrading enzymes (proteases) that are heat-stable. Added to laundry detergents, they allow clothes soiled with proteinaceous material to be washed at high temperatures, since they are still able to degrade proteins. As boiling an egg shows, normal proteins coagulate at 50 to 60 degrees, becoming inactive.

Microorganisms or their enzymes have been employed for thousands of years in biological processes such as brewing beer and making cheese. In the future they will be applied in much more specific and controlled ways, resulting in a refinement of the biotechnological processes now in use.

Protein Production

The production of insulin and the growth hormone somatostatin has already shown that proteins either difficult to prepare or obtainable in only small amounts, usually hormones, can be prepared in batch cultures of coli bacteria or yeast via gene insertion. The resulting insulin or somatostatin, even though produced in a microorganism, is identical to *human* insulin or somatostatin. Its physical properties and biological activities are indistinguishable from those of the naturally produced protein. The human insulin prepared today on a large scale is for the most part the product of genetic engineering. Although this approach represents an improvement and refinement from both a pharmaceutical and an industrial point of view, it is not fundamentally new. The production of some vaccines against viruses is novel, however. Many viruses, such as hepatitis B, have so few proteins of their own that the human organism cannot generate antibodies against the proteins produced by these viruses. Therefore, a viral DNA segment coding for a viral protein is cloned in coli bacteria* and the viral protein thereby produced is then used for immunization.

Research in the area of so-called endogenous protein hormones is making rapid progress. Most regulating hormones are proteins of great pharmacological interest. They are produced primarily in the hypophysis (pituitary gland) and are present in the body in only very small amounts. Genetic engineering most likely offers the only way to obtain them in reasonable amounts. Protein hormones appear to play an important role throughout the central nervous system; for example, they mediate sensitivity to pain (endorphins) and are possibly involved in the transmission of electrical impulses. Without the methods of genetic engineering, neither basic research in this area nor pharmaceutical applications could witness further progress.

New Microorganisms

A coli bacterium genetically engineered to produce growth hormones is a microorganism with new properties and – depending on the extent of these new properties – it might even be considered a completely new organism. Two directions of development are possible. Either certain (bad) genes could be removed from the microor-

* Here, to clone means to insert a gene or a piece of DNA into a microorganism and then to grow the organism in culture; the gene is therefore present uniformly in the resulting culture.

ganism or new (desired) genes could be introduced into it. Pathogenic bacteria are usually harmful (pathogenic) because they produce poisons or toxins that, in even minute amounts, are severely detrimental to the human body. Examples include diphtheria toxin, cholera toxin, the toxin of botulinus (which causes botulism), and many others. It is possible, in principle, to remove the genes responsible for production of these toxins from the respective bacteria. The resulting bacteria are no longer dangerous and could be used for immunization. In fact, the principle of immunization with "weakened strains" has long been applied. However, these strains can now be produced in a more straightforward and specific way.

Extremely important is the construction of new organisms capable of degrading waste. In fact, most waste water is now clarified and purified with the help of microorganisms. As long as the settling ponds are properly aerated, these microorganisms feed on the sludge and ultimately decompose it; in other words, they gently "burn" it. Clearly, then, water purification has become a largely microbiological art. But there are some substances for which no bacteria are specialized; examples include chlorine-containing hydrocarbons, polyethylene, crude oil, asphalt sludge, and many more. Present research is aimed at finding specific microorganisms – usually pseudomonas species – genetically capable of degrading complex substances not found in nature.

These methods could have drawbacks, though none have been become apparent so far. Petroleum-consuming bacteria might start to grow in oil tanks or at least on lubricated surfaces or oiled machinery; indeed, they could conceivably gobble up the crude oil in oil wells. Bacteria capable of degrading plastic might start to nibble away at the plastic parts of automobiles as well as PVC floor coverings and other plastic objects in homes. Toxin-producing bacteria might escape into the environment and contaminate it. A frightening possibility is even the colonization of the human intestine by homone-producing coli bacteria, resulting in overproduction of hormones. Insulin-producing bacteria, for instance, could yield an enormous amount of insulin, which, if it entered the bloodstream, would lead to insulin shock.

All these technological developments constitute potential interference in the complicated equilibria of nature. At each step, therefore, their possible effects on the ecosystem in question – the human intestine, for example – must be ascertained beforehand. This is why strict laws govern the construction of new microorganisms. In Germany, for example, all experiments of this kind must be reported to the government and first receive approval before they are carried out.

Over the past fifteen years, experience with genetically modified microorganisms has shown that each change in the genetic makeup of a bacterium puts it under such a great disadvantage compared to the wild type that it is no longer able to survive "in the wild." For instance, coli bacteria are so finely adapted and equipped to carry out their vital functions and survive in the human intestine that even a 10 percent production of insulin, which in itself is useless to coli bacteria, represents an unbearable burden. Such strains exist only under artificial laboratory conditions, therefore, and cannot survive in the human gut. Life in the "free environment" is just as lethal for these bacteria as taking a walk on the moon would be for an astronaut unequipped with a space suit.

Nevertheless, this practical experience should not cause us to relax our guard, nor should a researcher's conscience rest once the requirement of legal notification is met. Each sally into unexplored scientific terrain brings with it an environmental risk. The discovery of X rays and penicillin and the development of the internal combustion engine, the television tube, winter wheat, and satellite communications carried risks for human society. The researcher's duty is to be the first to recognize such dangers and immediately sound a warning.

New Plants

The cultivation of new, higher-yield, more resistant – in short, better – varieties of plants dates back to the beginnings of human culture. Indeed, the word culture is derived from the Latin *colere* (to cultivate). However, what was once accomplished by crossing and breeding over many plant generations, guided by chance and subsequent selection, is now achievable by the new methods of genetic engineering. They allow a single feature to be introduced selectively and the classical goals of cultivation to be attained in a much simpler and faster way. These goals include higher yields, hardiness to frost, resistance to pests, decreased water requirement for cultivation in arid zones, and reduced perishability.

The modern methods of genetic engineering, however, are fundamentally different from previous methods of cultivation: they are not restricted to certain species. A gene – that is, the instructions specifying the production of a new protein – can be introduced into one species from a completely foreign species. In principle, wheat plants could become capable of generating proteins normally found in chicken eggs or cattle. Of course, the problem here is much more complex than that encountered in bacteria. Compared to unicellular organisms, plants are much more highly specialized. Moreover, a genetic trait like hardiness to frost is certainly not governed by a single gene, not even by just a few genes. Many genes would probably have to be modified. This means that we are once again confronted with complex network systems. They cannot be modified by simple causal techniques, which might indeed destroy them.

An important problem and the focus of work in many laboratories worldwide is the construction – I prefer not to use the word cultivate any longer – of plants capable of assimilating (fixing) nitrogen from the air. Most cultivated plants require nitrogen fertilizers, since they alone are incapable of obtaining from the air the nitrogen required to synthesize proteins and nucleic acids. Only a few species possess this capability. Legumes, for example, are associated symbiotically with nitrogen-fixing microorganisms, which colonize their roots forming small nodules. These nodules are useful since the bacteria within them supply the plant with nitrogen. The chain of enzymes catalyzing nitrogen fixation is coded for by a set of genes called the nif genes (nif is an abbreviation for *nitrogen fixation*). Much work is presently aimed at introducing these genes into wheat and corn. If this work is successful, these plants will no longer require nitrogen fertilizers.

Attaining this goal would open up boundless vistas – but also risks, those inherent to any technological development. In this case, as in others, the consequences are especially grave because of the breathtaking pace of technological development. The transformation of Europe from primeval forest to cultivated farmland took about 6000 years, from the beginning of the Neolithic age up to the late Middle Ages. The massive ensuing change in the environment could be coped with culturally, because mankind was able to adjust to an agricultural society over many generations. But how long will we have to adapt to the changes in our environment made possible by recombinant DNA technology?

I do not mean to imply that these new methods are bad. For a world that is increasingly interdependent and whose population is growing constantly – today, parts of the world population are starving – such methods are probably necessary and useful. At first glance they appear to be ethically neutral, but what we do with them is of ethical import. All stages of development, especially the first practical applications, must therefore be accompanied by an ongoing technological assessment of their consequences.

New Animals

The closer we approach man on the ladder of evolution, the more critical and thought-provoking our situation becomes. At the same time, the complexity increases in an unforeseeable way. All applications of genetic engineering to animals require that the "genetic operations" be performed at the egg-cell stage, that is, shortly after fertilization. Only in this way can the complete complement of genes and hence all future cells be modified uniformly. This requires that fertilization take place in vitro, that is, outside the body. In vitro fertilization no longer poses any difficulty in the case of animals. However, if the genetic operation is performed at a multicellular stage, the new or modified gene is inserted into only one cell or at most a few cells of, say, an embryonic organism consisting of hundreds of cells. Accordingly, the gene is active in only certain parts of the mature organism. This means that the corresponding protein will not be produced everywhere but only in a few tissues, such as bone marrow, the central nervous system, or connective tissue. A gene inserted in this way is certainly able to exert its effect, but it is not necessarily passed on. It all depends on whether the germ cells produced in the fully grown organism are derived from tissue containing this new gene.

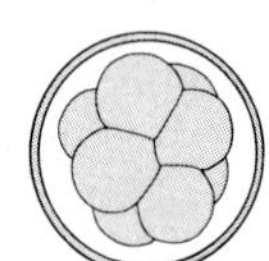

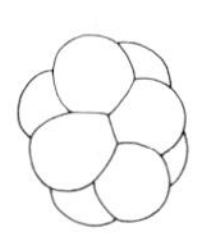

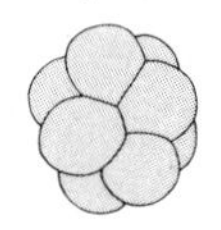

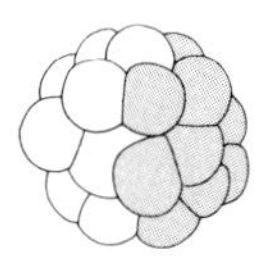

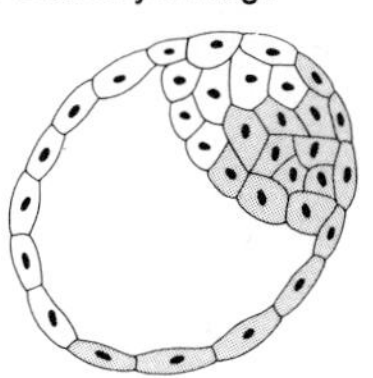

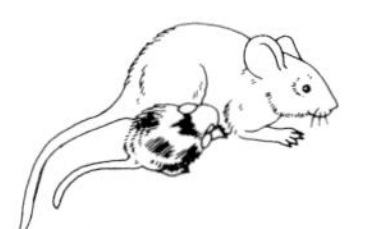

Fig. 3.6. A method for producing chimeric mice; two morulae with different genes for hair color are combined (the embryo is roughly at the eight-cell stage).

In a certain sense, the different tissues and organs of these chimerae* (Fig. 3.6) have different parents.

Figure 3.7 shows a goat obtained by fusing embryonic cells of a goat and a sheep – a "goatsheep," so to speak. This kind of grafting is carried out at a very early em-

* Chimera: a creature of Greek mythology with a lion's head, a goat's body, and a dragon's tail. In genetics, a chimera refers to an individual formed by grafting of two species.

Fig. 3.7. The sheepgoat.

bryonic stage. Complete integration seldom occurs at later stages – for example, in organ transplantation.

Are such violations of natural boundaries ethically permissible? We allow ourselves to slaughter animals and, after seriously weighing the justifications, to use them for some medical experiments. Crosses between different kinds of animals have been performed since antiquity. The mule is one example. Although it was produced by natural fertilization, it can still be called a chimera. I see no reason, then, why genetic and breeding experiments with animals should be ethically questionable.

Gene Therapy – What Is Medically Feasible in Man?

Genetic Analysis

Each application of gene therapy first requires a diagnosis. Diagnostic approaches have long been available to identify certain genetically based diseases, such as hemophilia, phenylketonuria (a metabolic disease in which toxic substances accumulate in

Biochemically defined genetic diseases

Name of the disease	Pathological change	Consequence
Sickle-cell anemia	Wrong amino acid in hemoglobin	Inadequate supply of oxygen, premature death
Phenylketonuria	Absence of an enzyme to degrade phenylalanine	Retardation
Chorea-Huntington diseae	Locomotion disorder	Genetically inherited St. Vitus's dance
Duchêne muscular dystrophy	Muscular dystrophy, Inadequate development of the motor end plates	Complete crippling
Hemophilia B	Lack of a blood coagulation factor	Lethal bleeding
Cystic fibrosis	Inadequate levels of growth hormone, inability to produce a sufficient amount of somatostatin	Dwarfism

the body, leading eventually to insanity or death), Tay-Sachs syndrome, and others. The table above lists a number of such diseases, all of them exactly defined biochemically.

Roughly 400 genetic diseases due to defects in single genes and about 2000 diseases involving defects in multiple genes are known. They all can be diagnosed medically, but only after the first indications of the disease have appeared. But the application of gene therapy requires an earlier and more precise diagnosis. The site of the gene in question has to be determined and its sequence elucidated. Potential approaches already exist and, in some cases, gene analyses are now possible. A novel technique, called *chromosome walking*, employs synthetic oligonucleotides to locate a specific gene on the chromosome; the gene is then cut out, analyzed, and unambiguously diagnosed[9]. In principle, this method is applicable to any cell in the body; even a skin cell of a fully grown human being can be subjected to genetic analysis.

What will be the consequences of such analyses? First of all, a much clearer understanding of the human body will emerge. Not only quick diagnosis of a specific disease but also of the genetic predisposition toward it will be possible. Doctors will be able to estimate the chances of acquiring certain diseases and to suggest appropriate countermeasures, such as a proper diet. The *positive* consequences, then, are readily apparent. But the *negative* consequences should not be underestimated. Before a genetic disease actually appears, the afflicted person might be denied access to health insurance. Companies would be able to assess the health risk of hiring a potential employee and possibly reject the candidate or limit the scope of his or her work in some way. Diagnostic procedures of this kind have to be in accord with the rights of the individual and with basic human rights. For example, should a person be denied life insurance if he or she refuses to undergo a genetic analysis?

The methods of genetic analysis take on particular importance when applied to genetic diagnosis of an embryo in the womb. In this procedure, called amniocentesis, a small number of embryo cells floating within the amniotic fluid are removed by aspiration. These suffice for chromosomal analysis. In the future, the procedure will probably also allow a complete genetic analysis of the embryo and an assessment of its chances of survival. Should corrective medical procedures then be permitted? What constitutes a medical indication for abortion? A severe mongolianism (Down's syndrome) associated with a short life expectancy and total idiocy? Hemophilia with a life subject to constant danger and dependence on expensive drugs, but otherwise normal? Albinism (deep red pupils and the lack of skin pigments) with an increased risk of sunburn and skin cancer? Or merely the absence of an X chromosome, one of the most common genetic defects, which occurs in roughly half the population and results in loss of the female sex? Men are heterozygous minus mutants with respect to parts of the X chromosome; they are genetically defective. Only women bear the full complement of genetic information in Homo sapiens (see chapter 1).

Prenatal diagnosis makes it possible in principle to bear a child of the desired sex, provided that the parents are willing to "abort" an embryo of the less desired sex. These practices already seem to have become widespread in some countries of eastern Asia. Because male babies are preferred, female embryos are often identified by chromosomal analysis and then aborted. As we will see in the following discussion on the embryo's rights as an individual (see p. 72), it is not permissible for a civilized and enlightened mankind, universally proclaiming the rights of the individual, to make an ethical distinction between an abortion based on an unfavorable prenatal genetic diagnosis and the deliberate abandonment of undesired or malformed newborn babies in the forest or mountains to perish.

The improved diagnostic possibilities and the necessarily ensuing medical questions have already led to very difficult ethical problems. The life-form known as man has not been given the license to intervene in the network of life.

Gene Implantation

A hereditary disease is often manifested in the absence of a particular substance. The body cannot synthesize a certain enzyme, for example, since the required genetic information is lacking. In one sense a biochemical organ is missing. Can an "organ transplant" alleviate this damage? This possibility is becoming ever more real. Sickle-cell anemia, due to an error in the hematopoietic (blood-forming) system, is one example. The red blood cells have an abnormal, sicklelike shape and cannot properly fulfill their function of oxygen transport. The resulting deficiency in the supply of oxygen to sensitive tissues leads to physical disorders and even premature death.

The precursors of red blood cells are produced in the bone marrow. If foreign bone marrow could be transplanted into a sickle-cell patient and were not rejected, then the patient would be cured – not of the hereditary disease, to be sure, because the disease would still be passed on, but cured as a person for his or her lifetime. The "genetically diseased" bone marrow would have to be removed and replaced by bone marrow from a healthy person. Provided that this new bone marrow grows normally, the patient can live a normal life. Genetically, of course, the red blood cells floating about in the patient's blood would be foreign, since their production would be governed by instructions introduced from outside the body. But this changes nothing in the fact that the person is cured. Admittedly, no major success has been scored in this area. Yet there can be no question as to its medical potential.

Gene Therapy Is a Violation of the Personal Rights of Future Human Beings

In order to modify or replace a gene in an organism, the nucleic acid in question has to be removed, chemically modified, and reinserted. This is now done on mice and what works for mice should in principle be possible for humans. This may sound somewhat cynical, but it is true insofar as the technical aspects are concerned. The prerequisite is that the germ cell be isolable in a test tube where it can be manipulated. Apparently, public opinion and the courts have largely accepted in vitro fertilization. In my opinion, however, it is a glaring example of human presumptuousness and I will discuss it shortly.

Let us agree that in certain limited situations – grave hereditary diseases, for example – the ethical prerequisites for manipulating the human gene are fulfilled and that such manipulation will become technically feasible within the next decade. Nucleic acid can be removed from cells and reinserted back into them; it is stable enough chemically to allow the manipulation of individual chains. The instruments are at hand (restriction enzymes, see p. 57) to excise defined fragments from the nucleic acid. Correspondingly, the gene fragments, once repaired, can be reinserted and rejoined or ligated to the nucleic acid. Certainly, what is now carried out with bacterial genes will soon be possible for human genes as well.

Still, the ethical question remains: Is all this permissible? Let us discuss this issue by way of a specific example. Say a family is threatened by hemophilia, as were the ruling houses of Europe at the beginning of this century. The female members of this family do not display hemophilia phenotypically. Although they put future generations at risk, they themselves are not in any danger. For the male members, on the other hand, the chances of acquiring the disease are very high, namely, 50 percent. Let us assume that it is possible to determine whether a fertilized ovum is male. In principle, this is now possible. For a 50 percent risk, it might seem justified to replace the gene in question. To accomplish this, further development of the fertilized ovum would have to be interrupted, for example, by cooling. The nucleic acid of the defective gene would be removed and replaced through hybridization. The repaired nucleic acid would then be reintroduced into the cell nucleus and, finally, the cell, capable once again of growth and division, would be put back into the mother's womb. There the child would be carried to full term. If the operation were a success, the child would be cured. I would like to emphasize that, while not possible at present, this scenario is by no means utopian.

But are such things permissible? In a time when the killing of an embryo by abortion – that is, the complete and final denial of its rights as an individual – is an acceptable practice, I see no convincing ethical argument against gene therapy, even if it carries the risk of killing the embryo or leading to birth defects. After all, there is a risk associated with any operation. Indeed, an operation is performed in the hope that a living being's chances of survival – and that also means its rights as an individual and its chance to develop freely – are enhanced. If this hope is justified and the expectation of success is reasonable, the operation must be performed. Any other course of action would be in violation of a physician's medical obligations.

I am now going to contradict what I just said. When a development is pushed to its limits it is not uncommon for paradoxes to arise. If examined critically, these paradoxes often reveal fallacies in the original assumptions. Close scrutiny of the extreme cases, then, brings the assumptions into better focus. Accordingly, the following argument is not acceptable: Genetic manipulation of the human embryo is permissible because even killing is sometimes permitted. No; in fact, the argument is just the opposite. The refinement of scientific approaches has made us aware that the characteristics of a future individual, who is worthy of our protection, are already fully developed at the embryonic stage. Not only the ability (or lack thereof) of blood to clot but all other abilities as well are present in essence: breathing, seeing, hearing, feeling, thinking, consciousness. The molecular magnification of the embryo made possible by recombinant DNA technology reveals that the embryo is in fact the individual itself, its "personhood" – merely dozing, as it were, in a nine-month slumber.

Although its body has not yet taken on its final shape, the embryo is already the overall ensemble of its future individual features as defined by its genes. We can change nothing in this ensemble without violating not only the virtual but also the real human rights of the individual, since the "molecular" magnification shows us without question that everything is latently present. In earlier times, when the state of biological knowledge was still crude, it may have been acceptable to regard the embryo as a hunk of flesh, unordered, soulless, mere happenstance, from which the individual developed only gradually, say, under the influence of the womb or through upbringing. This view is no longer justified. Not only is the individual latent in the fertilized ovum, it is actually present. This point of view is a compelling consequence of our current state of scientific knowledge. Of course, the straightforward conclusion is that interrupting pregnancy in any way is ethically reprehensible, since its goal is to kill an individual that already exists. A further conclusion is that any genetic manipulation of the germ cells or of the embryo is just as immoral as it would be to needlessly perform an operation on a mature person, either asleep or deeply anesthetized, without his or her consent.

The Limits of Gene Therapy – The Temptations of Knowledge[10–15]

Many, indeed most genetic characteristics, particularly those important for man, are not governed by a single gene. They are multigene effects. In living things, the combined action of several genes and the fine balance of their individual effects determine the characteristics of higher behavior in an extremely complex, interconnected way. The totality of genes, the genome, is a fundamentally complex network (see chapter 9)[16]. The loss of a single gene can, but need not, have a specific effect. It can lead to higher organization or to chaos. The system of genes is a finely tuned nonequilibrium involving the countless effects governed by the genes. It is therefore impossible to analyze this system in terms of causal relationships. What we call genetic diseases are merely the simplest genetic malfunctions, often involving only a single gene. When one considers the use of genetic engineering for the generation of positive characteristics such as intelligence, musicality, and linguistic talent, insoluble problems soon bar the way.

Why in Vitro Fertilization?

By the end of 1987, roughly 6000 children worldwide had been conceived outside the womb, in vitro. Without this artificial procedure, these test-tube babies would not have been born. These 6000 children are thus the product of an immense amount of medical research and skill as well as clinical care. And how many babies were aborted during this period? 600000 or 6 million or 60 million?

I fear that western civilization has already succumbed to the temptation of knowledge insofar as test-tube babies are concerned. Clearly, it is a hardship when potential parents learn that they are unable to conceive a child naturally. But why shouldn't they adopt children, perhaps from the Third World? Isn't it about time that we gradually free ourselves from the ingrained notion of "one's own flesh and blood" in a world that is growing ever smaller? Anyone who has ever visited the birth clinics in tropical Brazil, where foundlings are just waiting to be adopted, will surely have come to realize that the use of in vitro fertilization to produce one's own flesh and blood is, at the very least, an excess of civilization. I am of the opinion that today we must reconsider our traditional attitude toward heredity and adopt a more cosmopolitan behavior. Inherent in the desire for a test-tube baby, it seems to me, is much racist egoism, vestigial biological thinking that culminated not many decades ago in Fascsim and now finds its counterpart in the pretentiousness of our affluent society.

Genes and Society

Let us take a look once again at molecular recombinant DNA technology and its industrial applications. It has become customary in recent years to justify the high costs of research by emphasizing its social benefits, say, in the form of products and export surpluses. But this commercial justification can easily backfire. Scientists engaged in basic research, from whom one has come to expect direct economic benefits, are open to the accusation that their research has not yet earned a single penny. This situation is especially likely to occur when the research involves recombinant DNA technology, where basic research and its practical applications are tightly interwoven. What a misguided view of basic research!

Unfortunately, though, recombinant DNA technology has already borne a special kind of capitalistic fruit. There are a number of companies in which researchers, supported by public funds to run academic laboratories, are exploiting the fruits of their research on the side. This is a dangerously tempting form of fleecing. Here, only open and critical discussion will prevent the prostitution of science.

Obviously, basic research should not be confined to ivory towers. In the future, too, it will depend on collaboration with groups interested primarily in applications, usually industrial companies. There is nothing objectionable about this. The golden age of organic chemistry between 1880 and 1920 emerged from just such a collaboration, which, viewed historically, did not impair the freedom of research and teaching. The status of basic research was so unassailable then that researchers never relinquished their scientific freedom. The situation is different today, however. Many scientists have grown greedy for money (or perhaps they been forced to be so) and this greed is exploited by companies, that is, by our commercial system. I pale at the thought that cloned genes, hereditary characteristics in bottles, might one day be bought and sold. Acts of creation listed on the stock market! Wouldn't that represent the absolute victory of an exclusively profit-oriented system in which true science finds itself at a dead end?

The threat to science is not only commercial, however. In this century, researchers have increasingly prostituted their science to the powerful, in both East and West. The temptation, then, is not exclusively the product of capitalist or socialist systems. At least since the Second World War, the political aspects of science have become increasingly evident: radar, rockets, jet engines, nuclear energy, penicillin, computers, and space travel are developments of military value as well. During the last few decades, such developments have become of growing public concern. But they are still not subject to either suitable and sufficiently transparent forms of organization or internationally accepted ethical norms.

The scientist thus limps through time, like Hephaestus, both much admired and much despised, an odd stranger[17]. Brecht described this state of affairs in "The Life of Galileo." Even Archimedes built machines of war for the tyrants of Syracuse. A famous German scientist and Nobel Prize laureate helped organize gas warfare during the First World War. Napalm was invented by a Harvard professor, who tested it in the courtyard of the chemistry department. Countless scientists have been and still are active in military research. Many research results, even seemingly harmless ones, are of potential use for military purposes, for totalitarian regimes, for terror. How could science, which set out in the 18th century to improve the lot of man, arrive at this unanticipated and horrible end?

Not Everything Is Biology – Biology Is Not Everything[18,19]

The Relation between Biological and Intellectual Information – Some Simple Arithmetic

How important can recombinant DNA technology and genetic engineering become? The genetic information in the nucleus of a human cell contains 10^9 bits of information inscribed in the various genes. Each year, on the other hand, the human intellect produces roughly 10^{18} bits of information, stored in lecture manuscripts, libraries, journals, and electronic media. We thus produce a billion times more information per year than what is stored in our genes and pass this information along to the next generation. Even if only one percent of this information were "important," the amount of intellectual information would be ten million times greater than the amount of information present in our genes. In short, intellectual evolution has overtaken biological evolution in importance; for us the latter has reached its end. It has become unimportant, completely negligible in comparison with the development of the human intellect. Is it at all reasonable to manipulate the biological part of our existence, the genes, now that our intelligence is able to furnish a large number of technical solutions to problems? Why repair a gene responsible for diabetes? It is much better and easier to develop insulin therapy. Why repair a gene responsible for nearsightedness. It is better and simpler to wear eyeglasses or contact lenses. In many respects, Homo sapiens is a deletion mutant – with respect to vitamins, for example. But we have learned nonetheless to nourish ourselves *sensibly* in order to avoid scurvy.

It is tempting, of course, for us to think that most problems have technical solutions. This dogma, a product of the dazzling successes of the scientific age, is mistaken and very dangerous. It leads us to overestimate science on the one hand and to condemn it unjustly on the other. Instead, we should take stock of our intellectual strengths, our power of judgment, and our moral values, and we should cease to shift all responsibility to technologically oriented science.

Our genetic machinery, the genetic information together with the "apparatus" for protein biosynthesis, is a necessary, indeed a wonderful instrumentarium. But we should not be captives to our fascination, for we are more – indeed, much more – than our biology alone. Although a Mozart piano sonata cannot be played and appreciated without music and a piano, it is more than just musical notation and an instrument used to produce certain tones. Initially in the mind of the composer, it is now in the head and fingers of the pianist, ready to be performed as often as desired, but never the same.

Let me quote Mozart himself. Mozart was known to compose most of his compositions in his head, to hear them "performed" there, and then to write them down while talking with friends, for example. A symphony or piano concerto was complete and existed in its entirety before it was written down on paper and played on an instrument: "For instance, while traveling in a carriage, or after a good meal, during a stroll, and at night when I cannot sleep, my best ideas pour out. Those that please me I retain in my head and even hum to myself – at least I have been told so by others. Once I have recorded the first one firmly in mind, the others follow, a lump being required to make a vol-au-vent, arranged according to counterpoint, the sound of the various instruments, etc. My soul warms it up if I am not disturbed; it grows ever larger and I spread it out ever wider and brighter. Indeed, the thing is nearly finished in my head, even if it is long, so afterwards I am able to view it at a glance like a beautiful picture or a handsome person. I do not hear it sequentially as it will later present itself to the imagination, but all at once, together. What a feast! Everything, the discovery and the creation, happens within me like a beautiful, deep dream. But to hear everything at once, that is the best of all."[20]

What Responses Does the Challenge of Modern Science Evoke?

Response 1: Romantic Denial
The profound anxiety, the criticism of technology in modern industrial societies, the fear of future progress, and an awareness of the limits of growth are justified and necessary. We are ultimately responsible for ourselves. But does this imply a "return to the stone age"? Many of the antitechnological ideologies and responses are nothing more than romantic denial. A priori, every technological advance is rejected and opposed by one side while the other merrily greets it as if it were harmless. Our immense affluence and the pluralism of our society have made possible new forms of asceticism, the simple life. But we should keep in mind that those who seek new personal and group life styles, no matter how nice such people may be, are basically taking advantage of affluence; they are exploiting the infrastructure of a high standard of living. The new alternative life styles have an important and necessary function in our society, so overflowing with affluence; to alert us, to admonish us that life should return to basic values and human measure. Our society suffers increasingly from spiritual impoverishment. Hardly any forms of communal life remain, since structures like the family are breaking apart, without anything new to replace them. It is natural for people in the society to seek alternatives; indeed, the seekers themselves constitute an alternative. Still, it will never be possible to organize our life exactly as they would like to imagine. I see a danger in a romantic movement. The loss of religiosity and the end of all earlier forms of human solidarity, such as those found in villages, families, and nation states, without their replacement by new forms of solidarity, results in a vacuum in which "green" can easily become an ersatz religion. It might come to pass that the sorcerer's apprentice cries out for the great master, the guru, the powerful dictator. Political irrationality always carries the risk of dogmatism and the use of force.

Criticism of technology does not release the layman from the obligation of constantly seeking to acquire a core of scientific knowledge. Unfortunately, though, there is a huge information gap between science and the public. Scientists themselves are partly to blame for this state of affairs, since they often fail to present the results of their research in understandable form – out of laziness or the lack of time, because of an outmoded elitism, or simply because they are unable to do so or too shy. Yet, equally often, critical laymen manifest a learning block when it comes to science. In extreme cases, they even take pride in their know-nothingism. Such people are outsiders, however, and no longer have a right to criticize modern developments. It is a social duty to learn about the natural sciences. I do not mean to imply that laymen should acquire the knowledge of experts. But they should consult experts. Since antiquity, action and speech in politics and public life have required men and women to have a basic knowledge of history, without necessarily becoming historians themselves. Similarly, a basic knowledge of science is essential today. We need an enlightened public in order to surmount future problems.

Response 2: Reaction

The reactionary answer to the problem at hand is as follows: Science is a self-regulating system, which usually arrives at the optimal outcome by a process of supply and demand, an international competition of ideas and experiments, a natural interaction between science and society. Although this model of science has long been obsolete, at least since the work of Jürgen Habermas[21], it is still one of the most popular in many minds. Theory has advanced far beyond this vulgar positivism, but our mental habits have still not caught up. The model of a system evolving according to Darwinian laws, one that is self-optimizing, is not appropriate to science. If it were, science would quickly fall into the hands of technocrats. Helpless, passive acceptance of technocratic chauvinism is fraught with peril, but, unfortunately, it is an all-too-common answer to the still unsolved and ever-increasing problems caused by science and technology.

Response 3: Utopia as the Golden Mean

An iconoclastic attitude toward technology, biotechnology, and genetic research is unacceptable. On the other hand, a reactionary acceptance of the unbridled exploitation of science would have catastrophic consequences. Is there a middle course?

The middle course – the golden mean, as it were – is to remain alert and critical but, at the same time, to seek compromises, to strive to acquire as comprehensive a knowledge as possible without being seduced into putting it all to use, to reexamine the proven wisdom acquired during the thousands of years that man and nature have coexisted without succumbing to reactionary conservatism, to be wary of simple solutions without becoming indecisive[22]. The middle course is the most difficult. The wise fool who proposes it will quickly find himself falling between two stools. Yet Homo sapiens has ceased being a mere biological creature for more than 100000 years. He is reducible to biology only by deleting sapiens. Only if we consciously avoid this reduction and distortion by every possible means will we escape the fate of "biologistic" human monsters.

Like art, science is above all a cultural asset, an expression of the human intellect and its creative fantasy; it is so important as a cultural asset that it is even protected in, for example, the German constitution. Of course, science and its technological applications can be used to earn money, to build up industry, to influence capital markets. But science itself should not be conducted according to the laws of capital markets or the principle of utility (*Usura*). The results of genetic research, our insights into the structure of living things, should not be marketed according to the principle of utility.

Ezra Pound

With Usura[23]

With usura hath no man a house of good stone
each block cut smooth and well fitting
that design might cover their face,
with usura
hath no man a painted paradise on his church wall
harpes et luz
or where virgin receiveth message
and halo projects from incision,
with usura
seeth no man Gonzaga his heirs and his concubines
no picture is made to endure nor to live with
but it is made to sell and sell quickly
with usura, sin against nature,
is thy bread ever more of stale rags
is thy bread dry as paper,
with no mountain wheat, no strong flour
with usura the line grows thick
with usura is no clear demarcation
and no man can find site for his dwelling.
Stone cutter is kept from his stone
weaver is kept from his loom
WITH USURA
wool comes not to market
sheep bringeth no gain with usura
Usura is a murrain, usura
blunteth the needle in the maid's hand
and stoppeth the spinner's cunning. Pietro Lombardo
came not by usura
Duccio came not by usura
nor Pier della Francesca; Zuan Bellin' not by usura
nor was 'La Calunnia' painted.

Usury: A charge for the use of purchasing power, levied without regard to production; often without regard to the possibilities of production.

Came not by usura Angelico, came not Ambrogio Praedis
Came no church of cut stone signed: *Adamo me fecit.*
Not by usura St Trophime
Not by usura Saint Hilaire,
Usura rusteth the chisel
It rusteth the craft and the craftsman
It gnaweth the thread in the loom
None learneth to weave gold in her pattern;
Azure hath a canker by usura; cramoisi is unbroidered
Emerald findeth no Memling
Usura slayeth the child in the womb
It stayeth the young man's courting
It hath brought palsey to bed, lyeth
between the young bride and her bridegroom
CONTRA NATURAM
They have brought whores for Eleusis
Corpses are set to banquet
at behest of usura.

4. Evolution – Phylogenetic Trees and Lightning

A dialogue between Georg Christoph Lichtenberg and Albert Einstein on causality, fulgurations, and whether the world is determinable[1]

LICHTENBERG: *The question of whether one should engage in philosophy by oneself deserves the same answer, I suppose, as the similar question, Should one shave oneself? If asked, I would answer: If one is capable of doing so, it is a splendid thing. I always think that one should learn to shave oneself, but by no means should one start at the throat.*

But now to the question at hand: *the only thing we are certain of,* my dear colleague, *is matter and the Newtonian laws describing its motion. For us physicists, then, the only philosophy is materialism. Of course, philosophy is a risky enterprise and it is very easy to cut one's throat. Be alert, observe the essentials, measure, and compare – this is the whole of philosophy.*

EINSTEIN: Your view of the role of philosophy is quite interesting, my dear colleague.

What you call materialism simply means to view things in the light of causality. This way of looking at things only provides an answer to the question "How?" but never the question "What for?" or "Why?" No principle of utility and no principle of natural selection can solve this dilemma. However, when somebody asks why we should help each other, make each other's lives enjoyable, produce beautiful music, have sublime thoughts, then the only answer is, If you don't know it intuitively, nobody can explain it to you. Without this innate sense, we are nothing and we would be better off dead. If an underlying reason is sought by attempting to prove that these things help to preserve and promote the existence of the human species, then the question "What for?" is all the more pertinent and a "scientific" answer all the more hopeless. To proceed scientifically at all costs, we could try to reduce our goals to as few as possible and to derive all the others from these. This will leave you cold, however.

LICHTENBERG: *But it is essential to proceed scientifically at all costs. My mind is astir with something that I cannot quite put into words, something like a deep mis-*

trust of all human knowledge except for mathematics and physics. We are compelled to think in terms of causes and explanations, because, without this effort, I see no other means of finding our way in the world. Admittedly, someone might hunt for weeks without shooting anything. But this much is certain, if he had stayed at home, he would likewise have shot nothing. Indeed, absolutely nothing. Outside, in the forest, the chances – no matter how small – are in his favor. Of course, we must first catch hold of something, apprehend and comprehend it. But is it, after all, what we believe it to be? At times I have even believed that seashells might have been formed in the mountains. It was not a definite belief, though, but rather a dark intimation of our inability, or at least of my own, to penetrate the secrets of nature.

EINSTEIN: *I cannot share your pessimistic view of knowledge. It is one of the most beautiful things in life to see interrelations clearly. Only in a very dark, nihilistic mood could you deny this.*

But please don't misunderstand me. I shall not be brought so far as to renounce strict causality until an attempt has been made to oppose it in a totally unprecedented way. The notion that an electron subject to a beam of light is capable of freely choosing the moment when it is ejected and its direction of motion is intolerable to me. If this were so, I would rather be a cobbler or a croupier in a casino than a physicist. The world as a huge roulette wheel is a depressing notion.

LICHTENBERG: To acquire knowledge of the world, we simply pick and choose cause and effect relations out of the multitude of events. Otherwise, no knowledge would be possible at all, as Kant has just written. In fact, perhaps reality is utter chaos. Clearly, our mind had good reason to develop a strategy of avoiding chaos. *Perhaps man should be called the "causality bear," in analogy to the ant bear. This is putting it somewhat strongly. The causality animal would be better. We sniff all around for cause and effect relations, because we feed on causality.*

EINSTEIN: *I am well aware that there is no causality with respect to observables. In my opinion, however, this does not imply that theory, too, has to be based on statistical laws. You believe in a God who plays dice and I in a world of objective being that is totally governed by laws, which I try to fathom by wild speculation. I firmly believe that somebody will find a more realistic way or a firmer foundation than I have arrived at so far. The great initial success of quantum theory cannot persuade me to believe in a fundamental game of dice, even though I am well aware that my younger colleagues attribute this attitude to hardening of the arteries. One day it will become clear which of these instinctive attitudes is correct.*

LICHTENBERG: *There is a big difference, dear colleague, between adhering to an old belief and rediscovering it anew. To still believe that the moon exerts an effect on plants is a sign of stupidity and superstition, but to believe it* again *is testimony to philosophy and reflection.* But please excuse me if I am getting a bit personal.

EINSTEIN: It's all right, Lichtenberg. But you are certainly not being overly considerate.

Admittedly, quantum mechanics is awe-inspiring. But my inner voice tells me that it is not the whole story. Theory explains a lot, but it hardly brings us closer to the secrets of the Great One. In any case, I am convinced that He does not play dice.

LICHTENBERG: *In the beginning is the observable, from which the physical laws are then derived. But what good are all conclusions drawn from experience? I don't deny that they are sometimes true. But aren't they just as often wrong? And isn't everything a game of chance? I think that reducing everything in nature to simple principles ultimately means to presuppose that such principles exist. And how does one prove this?*

EINSTEIN: Dear colleague, *I am unable to justify my attitude toward physics in a way that you will somehow find reasonable. I cannot seriously accept the fundamental randomness of nature, because this is incompatible with the basic assumption that physics should be capable of describing a reality in time and space without any mysterious action at a distance. But I am not firmly convinced that the theory of a continuous field can really do this. I am convinced, however, that we will one day arrive at a theory that relates things not by rules of probability but by conceivable facts. However, I cannot provide logical support for my conviction, but solely my little finger as a witness – that is, an authority that will be met with no respect whatsoever, except from my hand.*

LICHTENBERG: *Such are the myths of physicists.* You are astoundingly honest, my dear colleague. This is the way things really are: *We grow aware of certain ideas that are independent of us. All that we know is the existence of our own sensations, ideas, and thoughts. One should say, "It thinks," just as one says, "It lightninged." "Cogito" is already saying too much once it is translated as "I think." To presuppose the "I," to postulate it, is required for practical reasons. Thoughts and lightning, taken as processes, are quite similar; they proceed via branching. At each branch point a decision is made. But who makes it?* It *thinks,* it *lightninged.* Isn't this "it," after all, the God who plays dice? *By the way, isn't it highly peculiar that a lightning bolt, though moving at such speed, seldom or never goes in a straight line and is easily diverted. This reveals that, in general, the bolt does not move a great distance, but instead springs from one point to another nearby. Are the branch points particles of air? What does matter look like infinitely close up? Take the microscope, for example. It just serves to confuse us even more. And no matter how far we are able to gaze through the tube of our telescope, we still see suns, probably orbited by planets. The magnetic needle points to a similar phenomenon in the earth as well. Suppose one could pursue this further and if, within the tiniest grain of sand, specks of dust revolved around specks of dust in the same way, even though it seems as motionless as the fixed stars? There might even exist a being to whom our visible world resembles a glowing pile of sand. The Milky Way could be one organic part thereof. To what extent could one explain the vegetation in terms of this system?*

EINSTEIN: *It is probably just as difficult to understand why an avalanche is unleashed by a mere speck of dust and thereafter takes its specific course.*

LICHTENBERG: We will just have to learn to live with such indeterminisms, such basic uncertainties, Einstein. Between two points *there is only one straight line but an infinite number of curves. Therefore, when a body is in motion, one could bet a million to one that the line of motion will be a curve. And, for each curve, there is a focus. Because circular motion is maintained the longest in the world, as we observe in the planets, in their motions around their axes as well as around the sun and the principal planets, perhaps all motion in the world originates from this. Light seems to be the only exception. But since it is presumably heavy, it, too, could be bent in its motion.* Indeed, this emerges from your theory of relativity, Einstein.

EINSTEIN: You are right, Lichtenberg. *At long last the generalization of gravitation is now utterly convincing and unequivocal from a formal point of view – provided that the Lord has not chosen a totally different path beyond our imagination. The testing of the theory is unfortunately much too difficult for me. After all, man is but a poor creature.*

LICHTENBERG: You always seem to toy around with your physics, honorable sir, but that, after all, constitutes your charm. *I have exhausted myself in pondering the explanation of gravitation and the conjectures on the formation of crystals, say, as if they were the Revelation of St. John. One is free to believe as much of it as one likes or is able to.*

EINSTEIN: For once this has nothing to do with belief, my dear sir. *To avoid this vagueness, it is essential to apply mathematics. And even mathematics attains this goal only by becoming totally insubstantial under the scalpel of clarity. Vital content and clarity are antipodes. One gives way to the other.*

LICHTENBERG: This is precisely our problem, dear colleague. *Perhaps man is half mind and half matter, just as a polyp is half plant and half animal. The strangest creatures are always found at the boundaries.*

The Evolution of Species – Phylogenetic Trees

We have seen in chapter 1 that life is a dynamical equilibrium between order and decay. In chapter 2, examples from biochemistry were used to shed more light on this notion. In particular, proximity to chaos was found to be productive. Finally, in chapter 3, the highly complex science of genetic systems was introduced. Nonlinearity is an ever-recurring feature in the systems studied today by modern biology; they branch and display discontinuities. Such systems evolve.

The concept of evolution was clearly formulated and introduced into science by Darwin[2]. He was the first person to conceive of and construct a phylogenetic tree of species, even though it was far from complete. By introducing a true natural history for the first time and thus historicizing nature (compare chapter 9, p. 222), Darwin brought about a paradigm shift, the central feature of a scientific revolution as defined by Thomas Kuhn[3].

Nature, regarded until then as static and dormant, is unchanging only to the eye of the short-lived beholder. In reality, it is a dynamical process in which one thing develops from another. The system of nature evolves.

Darwin's theory is well-known. Under the pressure of natural selection, the genetic variants of a species best able to adapt to their physical environment survive and multiply. Species therefore drift ever further apart. The variation of species ultimately arises from point mutations in specific genes. Most of these mutations are disadvantageous and lead to extinction of the corresponding genetic variants. The few variants with advantageous changes propagate.

By comparison of genetic traits Darwin established phylogenetic relationships; paleontologists later reconstructed more ancient relationships from fossils. Biochemical methods of taxonomy, developed over the last fifty years, have refined these phylogenetic trees still further. Today the most reliable method is sequence analysis of nucleic acids. A specific gene – for example, that coding for hemoglobin or cytochrome *c* (one of the respiratory enzymes) or so-called 5S RNA – is isolated from a variety of species and the various sequences are then compared (see chapters 2 and 3). The further apart the evolutionary development of the individual species, the greater the difference in these base sequences. In fact, this difference is quantifiable numerically. Figure 4.1 shows the evolutionary tree of man, the classical representation of the evolutionary tree shared by all primates.

By way of comparison, Figure 4.2 shows the phylogenetic tree determined for higher forms of life from the base differences in the gene coding for the enzyme cytochrome *c*. Between present-day monkeys and man, for example, there is a difference of only a single nucleotide ($0.8+0.2 = 1$), the same as that between a horse and a donkey ($0.9+0.1 = 1$)[5].

In the same way, 5S RNA may be used to derive the evolutionary tree of the entire animal, plant, and bacteria world; this nucleic acid plays an important role in the biosynthesis of proteins and is present in all forms of life, from the most primitive organisms to man[4]. The interrelationships of the species can now be determined directly not only qualitatively but also quantitatively from the number of different base pairs. Accordingly, the validity of the Darwinian description of an evolving nature is no longer questionable. But is the number of base pairs the sole possible criterion? Are monkeys and man really as closely related as are horses and donkeys? Are there no other criteria besides those based on molecular biology?

Let us now turn to another question: What determines the direction of evolution? Or, alternatively, what criteria govern the branching of the evolutionary tree? The Darwinian theory superseded once and for all the vitalistic, teleological theories of creation, though this was not Darwin's intention. In their place, his theory set up an operational scheme involving chance and necessity[6].

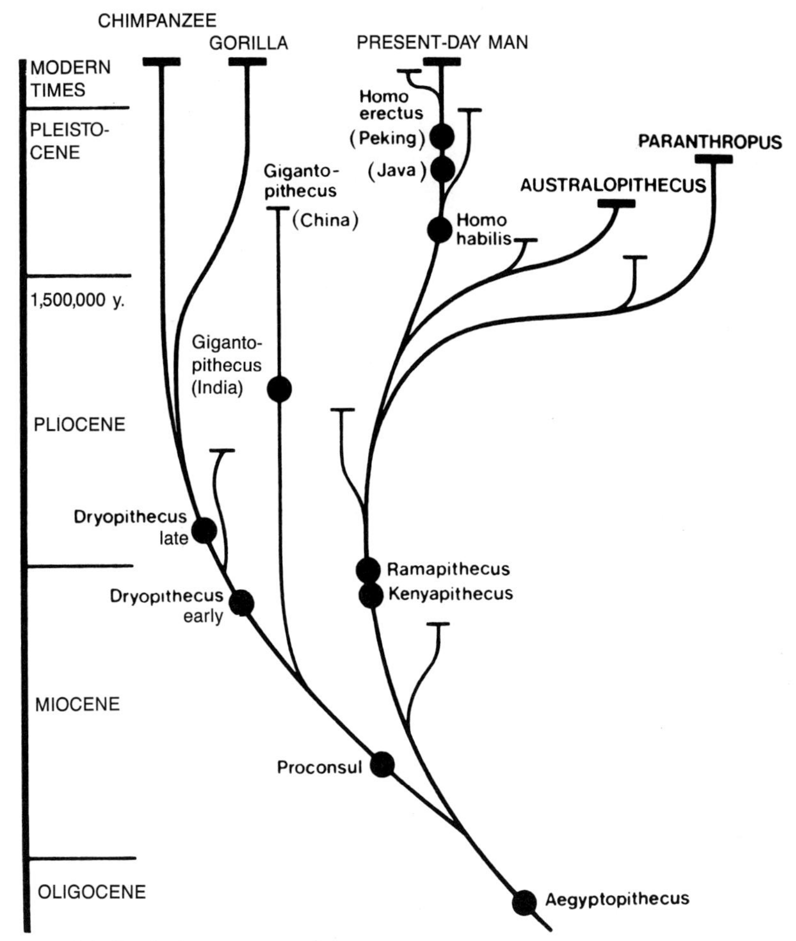

Fig. 4.1. Evolutionary tree of man.

Single point mutations or, as we now know, transposition of larger gene fragments via plasmids, a process called exon shuffling, result in a potentially wide range of genetic variations, which are subject to selection (see chapter 3). Each branch point thus represents an irreversible step. The genome changes; the new genetic traits, if adapted to a certain physical environment, are successfully expressed there – and only there. If they are not adapted to this environment, the organism cannot survive there and eventually dies out. Each branch point of evolution, then, corresponds to a jump, a discontinuity. Outwardly, evolution is manifested by a continuous system of living things; for that very reason it could be discovered by Darwin. In reality, though, evolution operates according to a mechanism that is discontinuous in both a physical and a mathematical sense. It occurs along an irreversible time scale. It was

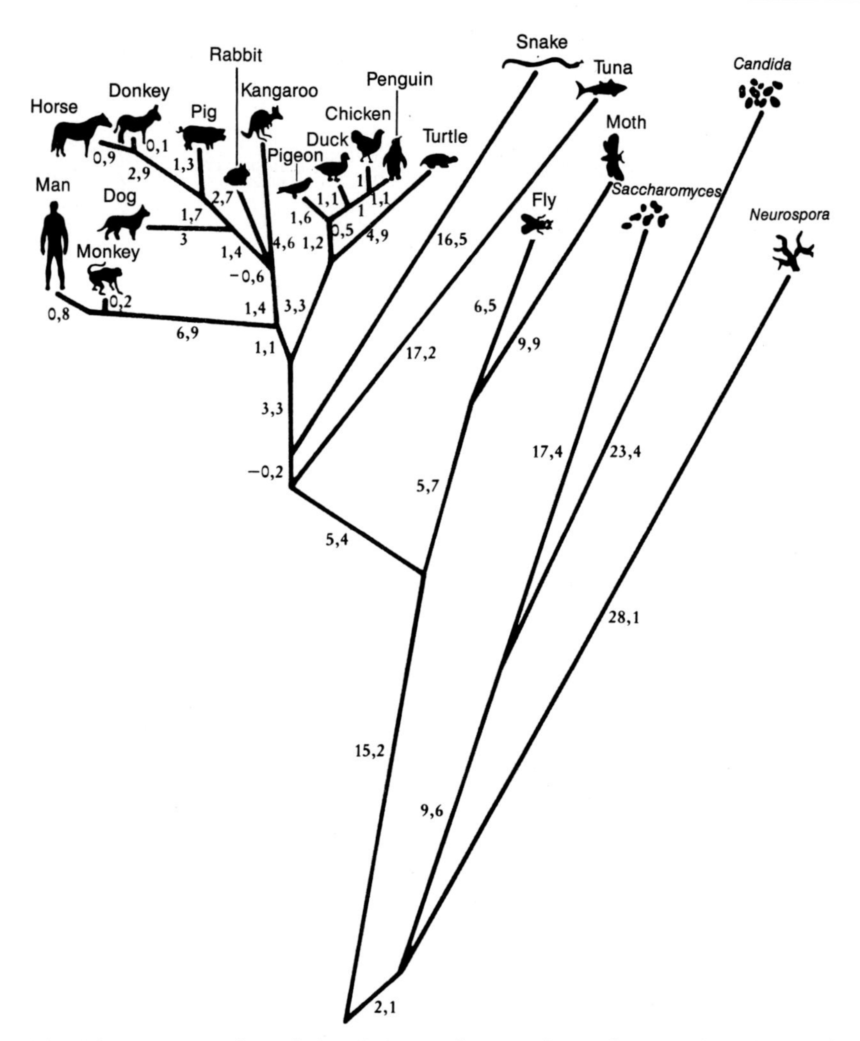

Fig. 4.2. Reconstruction of the phylogenetic tree of cytochrome *c* by comparative sequence analysis. The numbers on the branches are the minimum number of nucleotide substitutions in the DNA of the gene that are needed to explain the empirically determined differences in the amino acid sequence of cytochrome *c*. The resulting evolutionary tree of cytochrome *c* agrees with the macroscopic evolutionary tree reconstructed from paleontological findings, with only minor differences[5].

this realization that led to a radical shift in physical paradigm. Newton, treating time as a scalar quantity, assumed that it is reversible in principle: trajectories are retraceable, pendulums swing back and forth, planetary motion is periodic. Newton regarded instances of irreversible behavior as special cases that, at least in the context of 18th-century physics, could not be analyzed. Since Darwin, however, the realistic cases, those affecting our very life and existence, have come to be recognized as irreversible. Accordingly, Newtonian physics represents a special case. At first, this consequence went unnoticed. It was inconvenient, even annoying, and ran counter to positivism. Around 1890, Ludwig Boltzmann took up the idea of irreversibility (see chapter 8), but not until Lars Onsager and Ilya Prigogine did it become a generally accepted notion. This subject will be discussed later in greater detail.

Molecular Evolution – Eigen's Theory of Hypercycles[7]

Prior to the evolution of primitive unicellular organisms and later of all higher species – the sole focus of Darwin's studies – "molecular evolution" must have taken place, starting from simple, unordered molecules and proceeding on to the structure of the cell itself. The missing links lie somewhere between the Big Bang, which was followed by the formation of galactic systems, the solar system, and the earth ("unformed, and void"), and the emergence of the first living cell. They will probably never be found, at least not in the literal sense of classical paleontology. Still, there are two ways to approach the problem of how life originated: on the one hand, by devising experiments that imitate the conditions prevailing at that time and, on the other, by drawing theoretical conclusions from known molecular structures.

A number of now classical experiments have proven that the key substances making up living structures, namely, amino acids and nucleotides, form spontaneously in nature under certain conditions. Three billion years ago there may have existed a primordial soup, that is, puddles containing a large number of dissolved organic substances such as hydrogen cyanide, aldehydes, amines, and simple heterocyclic compounds. Electrical discharges or volcanic phenomena could then have induced chemical reactions to take place, leading to the spontaneous formation of simple amino acids and nucleic acid bases. Experimental attempts to imitate the primordial conditions on the earth's surface have been carried out in many laboratories around the world. The substances produced in these experiments include the important amino acids, such as glycine and alanine, as well as the key nucleosides adenosine and guanosine. Many of these organic compounds have also been found in meteorites. Moreover, cyanides and carbon radicals have been identified in interstellar space. In principle, then, inanimate systems exhibit a "spontaneous organic chemistry." The question still remains, though, how these inanimate systems gave rise to self-reproducing and self-selecting groups of molecules and, ultimately, to life itself.

We have already seen in chapter 1 that pure chance alone is an inadequate explanation for the origin of life. The evolution of life on earth has proceeded much too rap-

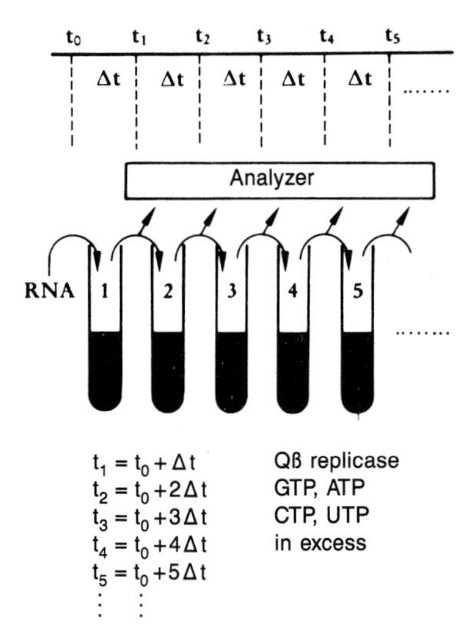

Fig. 4.3. Serial-transfer experiments on the replication of viral RNA. The technique is schematically shown on the left. RNA, especially RNA from simple bacteriophages such as Qß and MS2, undergoes replication in a medium containing the enzyme Qß replicase together with the four nucleoside triphosphates GTP, ATP, CTP, and UTP in excess. An aliquot is removed from the test tube after a fixed period of time (Δt). Part is analyzed, part is transferred to a new medium. This procedure is repeated at regular intervals Δt. The typical course of the experiment is shown schematically below. The rate of RNA synthesis increases in a stepwise fashion until, after one hundred or so transfers, it approaches a maximum value[9].

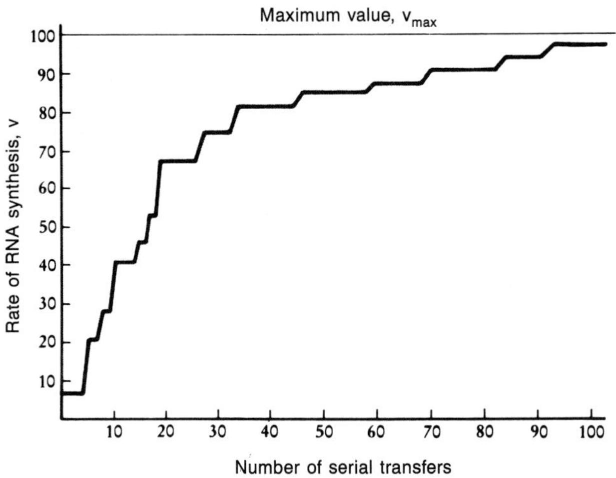

idly to be accounted for by simple random fluctuations. Instead, a mechanism of selection, operating from the very beginning, must have been responsible for choosing the "right" molecules and ensuring their "survival." Since there was no set of criteria for "right" and "wrong" – say, the ability to fly faster, to camouflage oneself better, to attract bees, etc. – only one criterion of selection could have existed at that time, namely, a higher rate of self-reproduction of an information-carrying macromolecule. It is now possible to carry out experiments on the self-reproduction of macromolecules under conditions corresponding to those existing at that time; bacteriophage Qß and its replicase, the enzyme responsible for copying the phage, may be used for this purpose[8,9].

The experiment is performed as shown in Figure 4.3. After roughly one hundred generations, the RNA formed is capable of replicating about fifteen times faster than

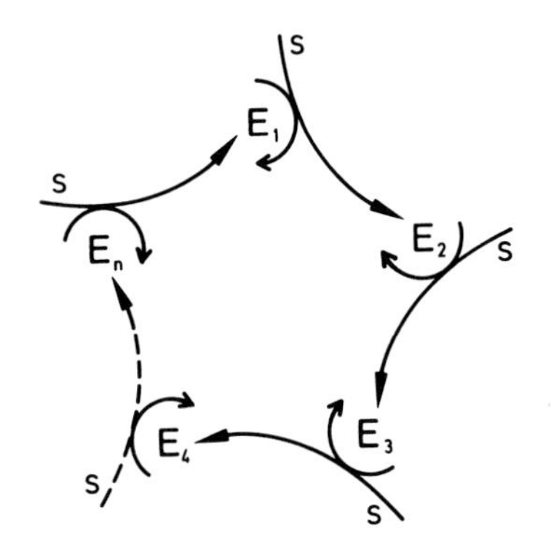

Fig. 4.4. Catalytic cycle in which several enzymes act together in such a way that continuous material transfer occurs. A detailed example is the citric acid cycle shown in Figure 1.12. By definition, a catalytic cycle is a closed system and hence an autocatalyst.

the original RNA. The system has "itself" carried out a selection to yield an ever-faster replicating RNA, but one with certain disadvantages. Its sequence has changed, it is shorter, and it has lost its ability to infect bacterial cells. Therefore, only *one* property of the system has improved: its ability to replicate faster. The experiment is biased and artificial. But it does demonstrate the following principle: molecular selection takes place in a replicating system under certain conditions without external interference[10].

A possible mechanism for the self-organization of molecules has been described by Manfred Eigen in his novel theory of hypercycles[7]. All biochemical reactions are catalyzed by enzymes. Several enzymes often act together, giving rise to a catalytic cycle, as we have already seen in chapter 2 for the citric acid cycle. A schematic catalytic cycle, which represents a higher level of catalytic organization, is shown in Figure 4.4. Not only are the individual enzymes in the cycle, E_1 to E_n, catalysts, but also the products formed in each catalytic reaction are catalysts for the next reaction. A catalytic hypercycle (Fig. 4.5), in turn, consists of several meshed catalytic cycles, each of which must fulfill two functions. First, each cycle must be self-replicating. Second, the products generated by one cycle must support the subsequent cycle.

This is shown more clearly in Figure 4.5. The information carriers I have two kinds of instructions, one for their own reproduction and the other for the translation into a second type of intermediate that functionally supports the next catalytic cycle.

Several conditions must be met in this system. First, the system must be capable of preserving its information and passing it on unchanged. The information for the sequence of a nucleic acid, for example, should not be rapidly lost. At the same time, the information has to be subject to change; the system must be capable of learning. In the language of nucleic acids, a change is called a mutation, the exchange of a nucleic acid building block. Accordingly, the sequence of nucleic acid building blocks must be able to tolerate change. A compromise has to be reached, then, between two properties: constancy and changeability. The nature of this compromise varies quite widely for the different forms of life, depending on their stage of evolution, the length of their genome, and the influence of their physical environment. We

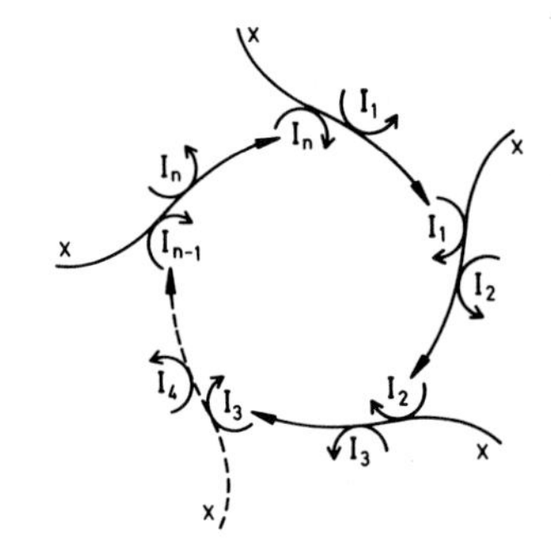

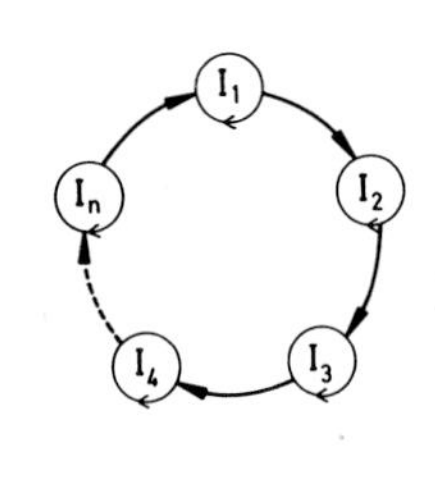

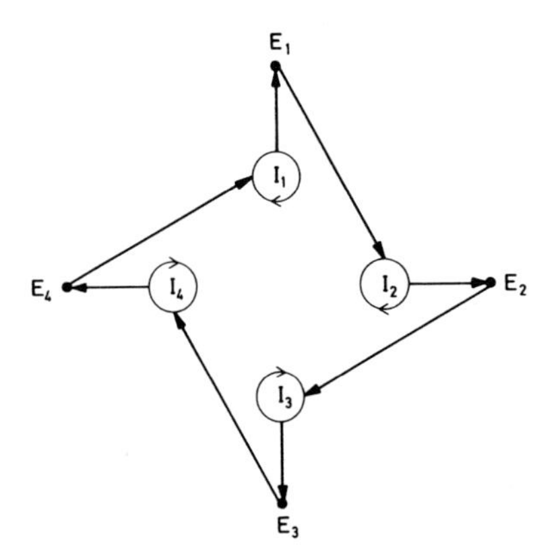

Fig. 4.5. Top: Catalytic hypercycle I (shown by the short curved arrow), a self-instructive entity for its own reproduction, provides catalytic support for the reproduction of the subsequent cycle (shown by the long curved arrow). Bottom: The information carrier I contains the information for its own reproduction and, in addition, the information for the translation of a second type of intermediate with optimal functional properties, usually an enzyme. This enzyme, generated by the information carrier, supports the activity of the subsequent information carrier.

have discussed this already in chapter 3. Phage Qß, a short information carrier of 3500 bases, changes quite rapidly. The accuracy of Qß replicase is about one error in 3500, so, as a rule of thumb, one exchange occurs each time the RNA genome is replicated. This exchange allows for the evolution of genetic characteristics. For man, this accuracy corresponds to one copying error in 10^9 to 10^{10} DNA base pairs.

Second, an evolving system, if it is to remain evolving, should not find itself at a dead end, that is, in an energy minimum, in a hole or valley (see chapter 1, p. 17), from which it cannot escape. After all, a salient feature of evolution is its unending course. Although they develop independently, the individual species are not totally independent of one another. Evolution is an interactive network and, apparently, it has always been so. Otherwise, the living world would consist of nothing but one dead end after another, that is, of fossils, animals at the ends of abstruse developments, already extinct or doomed to extinction – say, dinosaurs or giant deer with strange-looking antlers, creatures whose development has ceased because they no longer participate in the interactive network of evolution.

Eigen's theory of hypercycles, by taking these conditions into account, probably provides the best description of the mechanism of molecular evolution. It is impossible here to go into the extensive biochemical part and the captivating mathematical

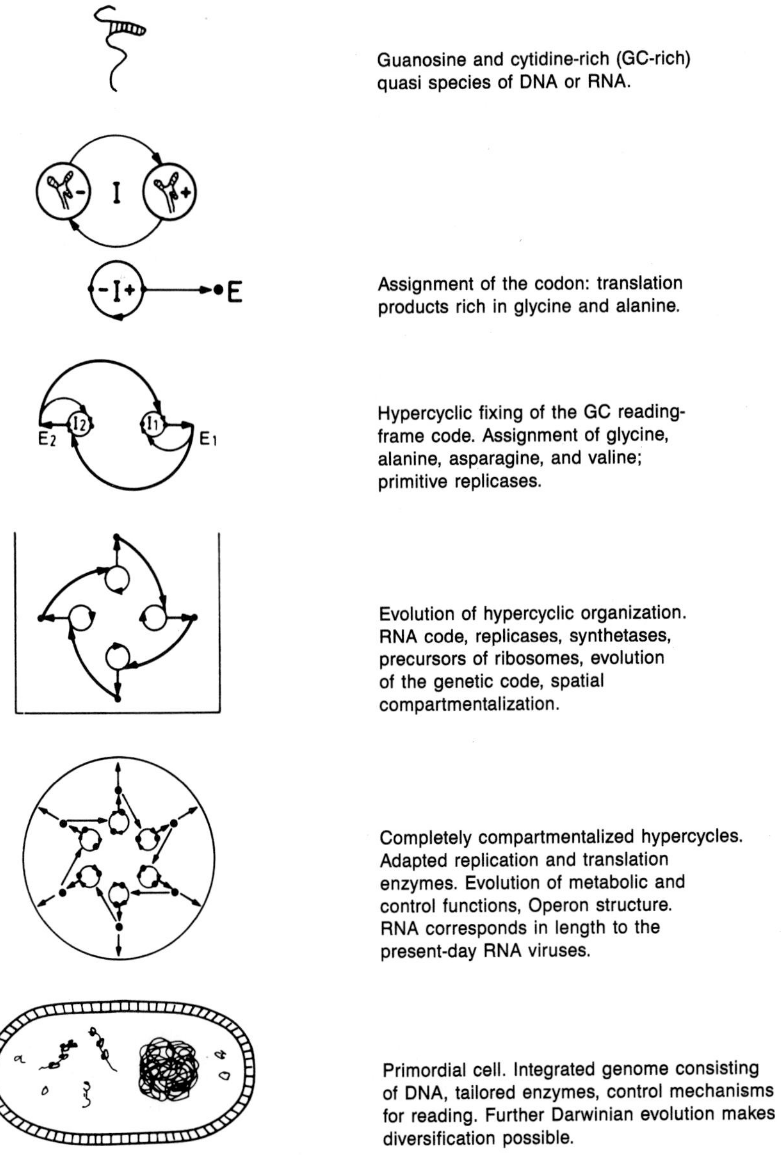

Fig. 4.6. Hypothetical scheme for evolution from a single macromolecule to the integrated structure of the cell according to the theory of hypercycles[7].

part of this theory in greater detail. The reader is referred to the monograph by Manfred Eigen and Peter Schuster[7]. A scheme for evolution from simple macromolecules to an integrated cell structure is summarized in Figure 4.6 in terms of the theory of hypercycles.

This theory is as important for understanding the mechanism of evolution as, say, quantum mechanics was for the physics of elementary particles. Heisenberg's quantum mechanics is a general method and form of representation for describing elementary physical processes, taking into account the statistical character and the uncertainty inherent in these processes. Eigen's theory of hypercycles is a general mathematical formulation of all possible evolving systems, taking into consideration the role of feedback in evolution. Quantum mechanics provides a mathematical description only; it is not "*the* explanation of physics." In the same sense, the theory of hypercycles is a mathematical method of description and not "*the* explanation of evolution."

The Evolutionary Theory of Knowledge – What Is Knowable?

The Two Sides of the Looking Glass

If living things arose through evolution, then their organs, particularly their sense organs also arose through evolution. What we hear, see, and feel has clearly proven useful to higher forms of life during the course of evolution. The sense organs of higher organisms include a receiver for certain wavelengths of the electromagnetic spectrum (the eye), for sound waves (the ear), for mechanical pressure and temperature differences (nerve structures in the skin mediating the sense of touch and temperature), and for certain chemicals (taste buds on the tongue and olfactory receptors in the nose). We perceive the world through these five senses; they paint our picture of the world. In a sense, the world is mirrored within us.

On the other hand, we cannot directly perceive radio waves, for example. We are able to do this only with receiving devices developed in the 20th century. Nor are we able to sense magnetic fields (although migratory birds and carrier pigeons do possess this ability). For this, we need a compass, invented in the 14th century. Nearly all living things employ the same basic physical mechanisms for perception: they analyze the light, sound, or chemicals in their environment with their sense organs. In some cases, the optical apparatus is more highly developed (birds of prey), in others the sense of smell or the ability to detect chemicals (dogs, insects, bacteria). The same basic pattern, though, seems to have proven useful to all living things, so they possess very similar decoding devices for sensing their environment. In this respect, the sense organs are almost as universal as the genetic code.

The human sense organs are also closely connected with the possibility of cognition or knowledge. Far from being arbitrary, the limits of knowledge are dictated by our sensory faculties, strongly augmented, of course, by our ability to think. Moreover, not only certain patterns of cognition but also certain patterns of behavior are genetically fixed. During the last few decades this has been shown by behavioral physiologists, notably Erich von Holst and Konrad Lorenz[11]. Since our evolutionary history may thus have set limits to and perhaps even programmed our cognitive possibilities, it is essential to investigate the a priori conditions governing human cognition. Kant did just that in his transcendental philosophy. In essence, the philosophical foundation laid in "The Critique of Pure Reason" still holds today. This work is not easy to read, however. Kant himself realized this and therefore wrote an introduction to his work that even laymen could understand. The "Prolegomena to any Future Metaphysic" begins with the sentence: "If one wants to constitute knowledge as a science, it is first necessary to distinguish what it does not have in common with anything else and what is thus unique to it; otherwise, the boundaries of all sciences overlap and none of them can be thoroughly treated in accordance with its nature."[12]

In behavioral research, Konrad Lorenz has carefully made this distinction and has also demonstrated to what extent human behavior is constrained by and dependent on biological prerequisites. He has thereby shown that some of the philosophical conditions and imperatives of knowledge discussed in Kant's transcendental philosophy are based on what we now know about the physiology of sensory perception as well as evolutionary history. In a sense, he has stepped "through the looking glass" of our sense organs.

Evolution as Ideology: Evolutionism

Darwin's theory of the origin of species represents a pragmatic attempt to explain the diversity of life in terms of natural history. Equally a part of life – and hence of evolution – are sensory impressions, behavior, and, not least, human thinking. This is one reason why a positivistic way of thinking threatens to debase the concept of evolution, to reduce it to an ideology. What was originally a scientific and pragmatic model threatens to become a worldview, evolutionism. How far-reaching is an evolutionary theory of knowledge?

Behavioral physiologists acknowledge, of course, that some forms of cognitive behavior are acquired through evolution[13,14,15]. Many human traits and behavioral patterns have turned out to be inherited behavioral reflexes without any moral value. One example is the "life-saving reflex" of the male baboon, which Konrad Lorenz characterized and which could effectively obviate life-saving medals. However, any attempt to analyze the ethical behavior of human beings from a purely biological standpoint – in terms of an evolutionary behavioral theory and an evolutionary theory of knowledge – would rob humans of their very being and personality, their uniqueness as individuals, and their human dignity! It would also place in question human freedom as well as man's responsibility for exercising a measure of control over his desires and emotions. Positivistic evolutionism completely overlooks the fact that we humans express mere *vestiges* of instinctive behavior. As the first to be set free after creation (Herder), we have been given "spirit." The cultural evolution thereby made possible is still occurring, also with respect to morals and ethics (see chapter 9). So, indeed, medals are still there to be won.

The prerequisites of scientific research are constantly being forgotten. Scientific research cannot yield moral values and cannot teach us the meaning of life. Naturally, sense organs, behavioral genes, and hormones play an important role in laying down a basic pattern, an underlying matrix, as I have already pointed out in chapter 3. But to reduce man to these alone would mean to dehumanize him.

Human "personhood," the self-responsible individual, moral values – these find no place in a worldview of evolutionism either as an "intermediate stage" (Carsten Bresch, chapter 7, reference 10) or as the "crown of creation," because they are ruled out at the very beginning of any scientific inquiry. In an equilibrium system in which, say, the quantity *b* is not present in the initial equation, but only *a, c, d, e, f*. . .*x, y,* and so forth, the quantity *b* can never appear in the final result, no matter how complicated the calculations might be.

No one is prevented, of course, from regarding the magnificent system of evolution as an exhaustive explanation of the world simply on the basis of its intellectual organization and rigor, from insisting on it, from deriving the meaning of life from it, just as people sometimes feel awe on viewing the star-filled heavens. But even Kant speaks of the "star-filled heavens above you *and* the moral law within you. . ."! The moral law must be included as the quintessentially human element. And this makes no statement whatsoever about its cultural and evolutionary origin.

A worldview without this "human element" is one of the mortal dangers confronting humanity. To view biology as an end in itself, to make it into an absolute, means to extend the mechanism of evolution to all areas of human life: it means the law of the jungle, social Darwinism, total genetic manipulation, and, ultimately, racism and the genocidal murder of human beings presumed to be inferior.

Phylogenetic Trees with Feedback for Fine-Tuning

Evolving systems with feedback controls and amplification, such as the Eigen hypercycles, are involved in the regulation of all living processes. Indeed, they constitute a principle of life itself. Several examples from biology will now be discussed.

The Ontogenesis of Nematodes

The nematode *Caenorhabditis elegans*, a small roundworm, is about one millimeter long and consists of only 1000 somatic cells[16]. Its anatomy has been elucidated in detail, cell by cell. There are 95 muscle cells arranged along the body, 102 cells in the nervous system, and 143 gonadal cells. Also known is the ontogenetic tree for all these cells, starting from the fertilized egg cell. The organism is formed through cell division according to a precisely defined scheme (Fig. 4.7). Feedback from neighboring cells directs and controls this process. It is an irreversible, discontinuous, ontogenetic process. At each branch point, something new arises – in this case, a new kind of cell.

Blood Clotting

Blood clotting involves the wall of the blood vessel, the blood platelets, and the blood-clotting factors present in the plasma and extravascular fluid. It is finely and precisely regulated by an intricate hierarchical system. This regulation is essential since the blood should clot only under certain conditions, namely, when an injury has occurred. Otherwise, a blood clot or thrombus could develop in an artery, possibly leading to an embolism. The bleeding of wounds, on the other hand, has to be stopped, or else an organism would bleed to death. The dynamical system of blood clotting operates along a narrow ridge with the danger of thrombosis on the one side and that of bleeding to death on the other. Fine regulation of this intricate system is made possible through a cascade of enzymes, which interact and control one another. This is shown in Figure 4.8.

Generation of Antibodies by Clonal Selection

The most important and characteristic property of the immune system is its ability both to respond in a highly specific way and to "detoxify" the millions of foreign antigens, foreign substances, and foreign cells that invade the body. Antibodies are presumed to arise by clonal selection and subsequent clonal growth. Consistent with this assumption, there are many different lymphocytes, each of which, at a certain point in its development, is programmed to recognize a specific antigen that has yet to see the light of day, that is, a hypothetical chemical structure. This huge diversity of antibodies, as many as 10^8 different ones, is generated by somatic mutation from a small number of basic structures. The immune system therefore requires just a few cells to guard against a possible invasion.

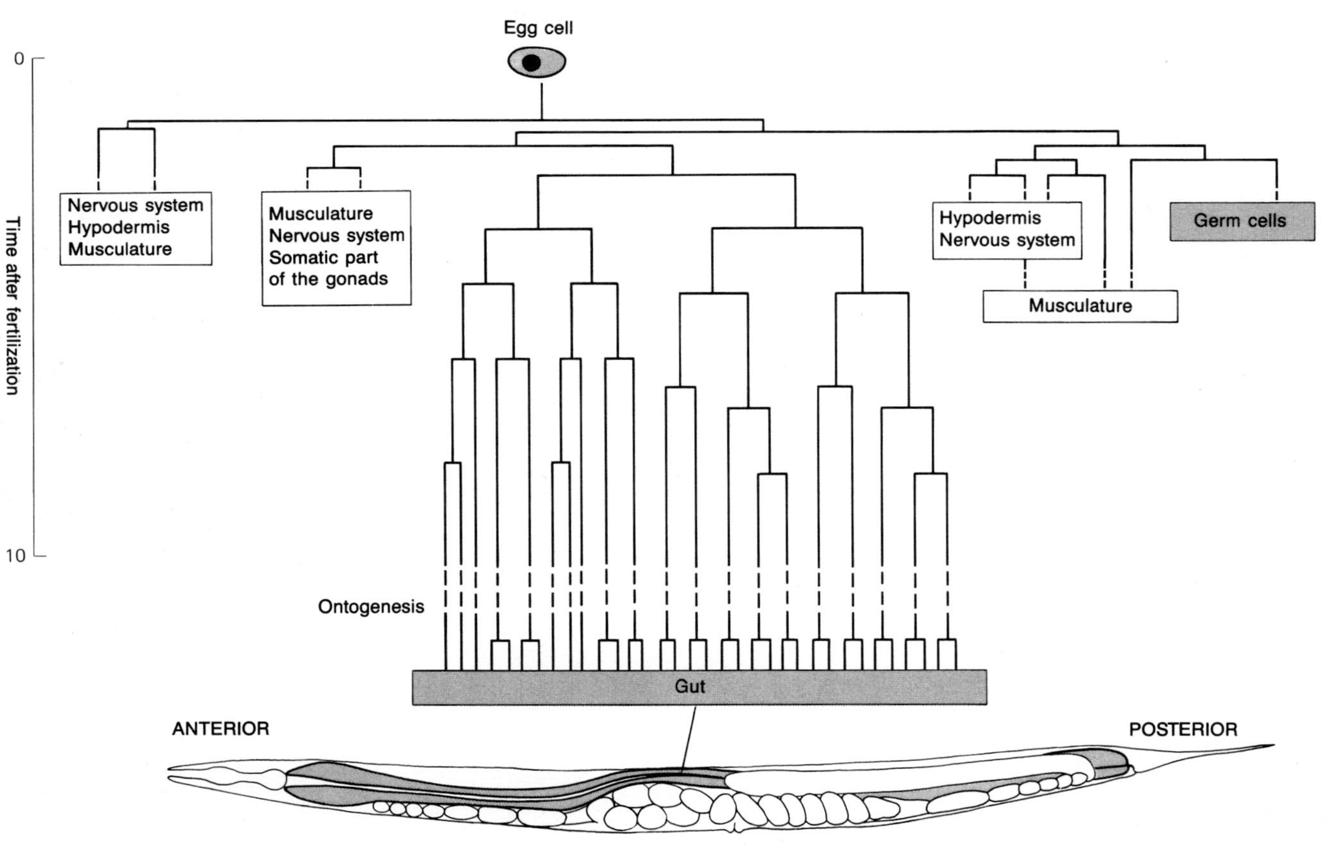

Fig. 4.7. Ontogenetic tree of the cells forming the gut of *Caenorhabditis elegans.* The egg cell (top) is drawn on the same scale as the adult animal (bottom).

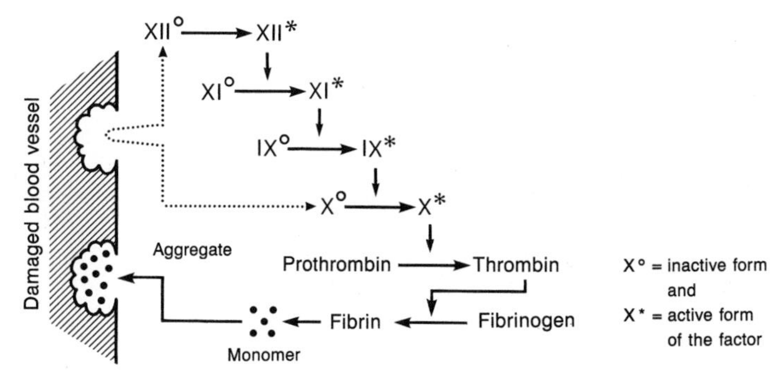

Fig. 4.8. Scheme of the blood-clotting cascade. The last step is the conversion of fibrinogen to fibrin.

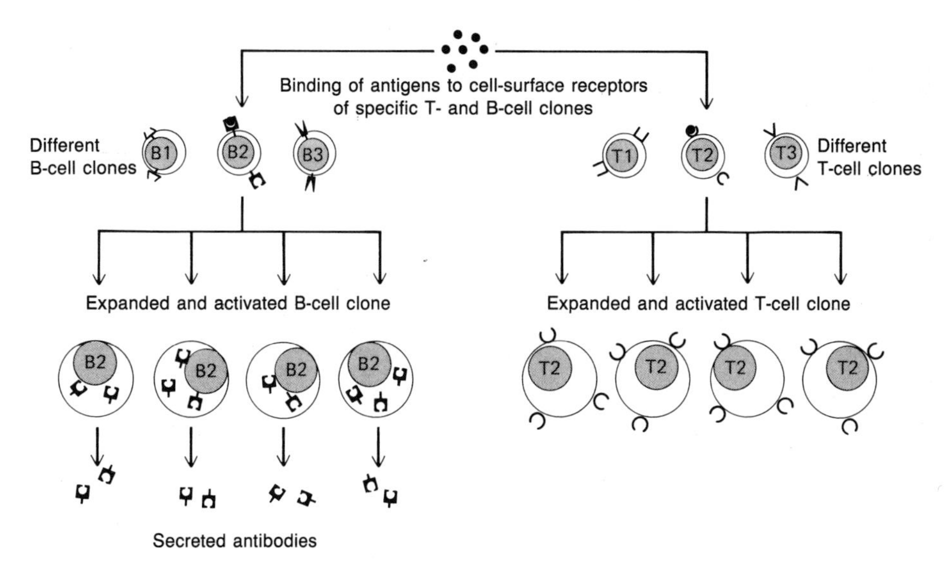

Fig. 4.9. Schematic diagram of the clonal theory of selection. An antigen activates only those T- and B-cell clones that are predisposed to respond to this antigen. The immune system is thought to consist of millions of lymphocyte clones, only a few hundred of which are activated by a specific antigen.

If one of the many conceivable antigens indeed penetrates the body, it is recognized by some of the cells, perhaps several hundred, and becomes bound to their receptors (see Fig. 4.9). This binding of antigen to receptor activates the cell and induces it to mature and multiply; a specific multiplication cascade is unleashed. The foreign antigen thus stimulates only those cells that possess complementary, that is, antigen-specific receptors, and hence already bear the imprint of the antigen. In other words, the immune response is antigen-specific. This process is illustrated in Figure 4.9.

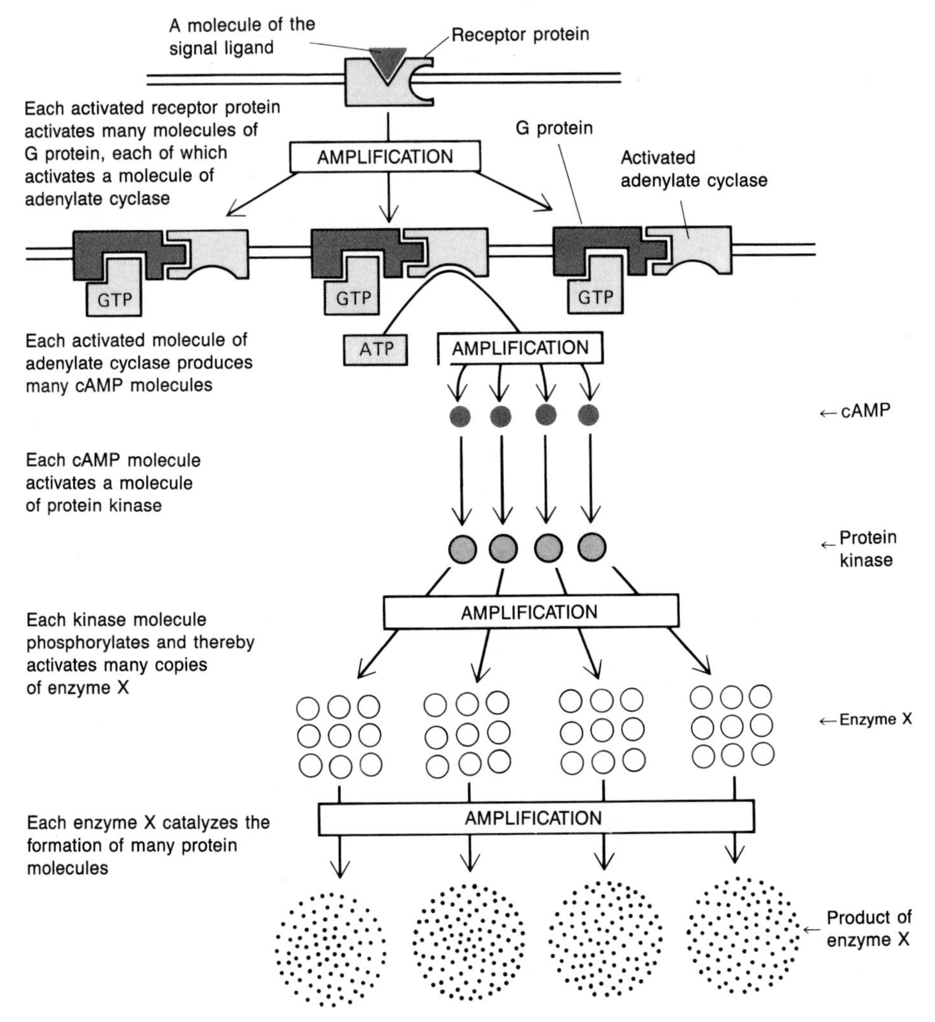

Fig. 4.10. Activation cascade induced by a hormone molecule. The original signal is amplified not only by transformation of the extracellular signal into an intracellular signal but also by coupled activation of adenylate cyclase by the cell-surface receptors.

Amplification of Enzymes by Activation Cascades

Many intracellular processes involving control and amplification proceed via branching activation cascades. All hormonal effects are due to processes of this kind. A small effect – the hormone – acts as a signal to induce a much larger effect. Often, so-called cyclic AMP, a universal intracellular amplifier molecule, plays an important

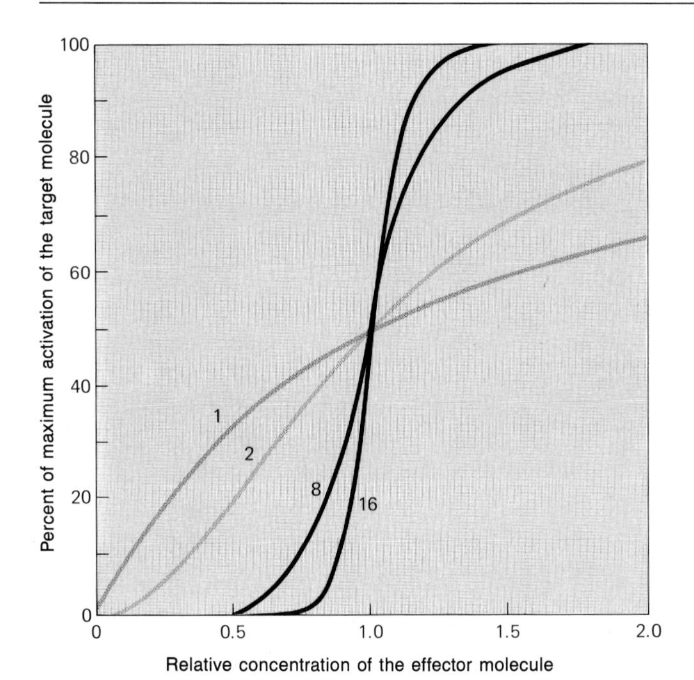

Fig. 4.11. Diagram illustrating how activation curves show an increasingly "all or none" behavior as more and more effector molecules need to bind simultaneously to a receptor molecule in order to activate it. The curves shown are those expected if 1, 2, 8, and 16 effector molecules bind simultaneously.

role in these processes (Fig. 4.10), regulating them and switching them on or off. The characteristic curves describing this control resemble those of systems involving electronic flip-flops; the processes are switched on or off according to an "all or none" principle (Fig. 4.11). An example is provided by the regulation of blood sugar level or glycogen degradation, a cascade system controlled at several stages (Fig. 4.12).

Another cascade system has already been mentioned, namely, the selection cascade used to discriminate between the amino acids isoleucine and valine (see chapter 2). All these systems have one feature in common: they are branched, nonlinear processes that are subject to feedback and proceed far from thermodynamic equilibrium. They are thus capable of achieving both high selectivity and high specificity.

Trees and Lightning

Lichtenberg himself carried out many fundamental experiments in electrophysics. He is the inventor of electrostatic printing, used today mostly in xerography. He was the

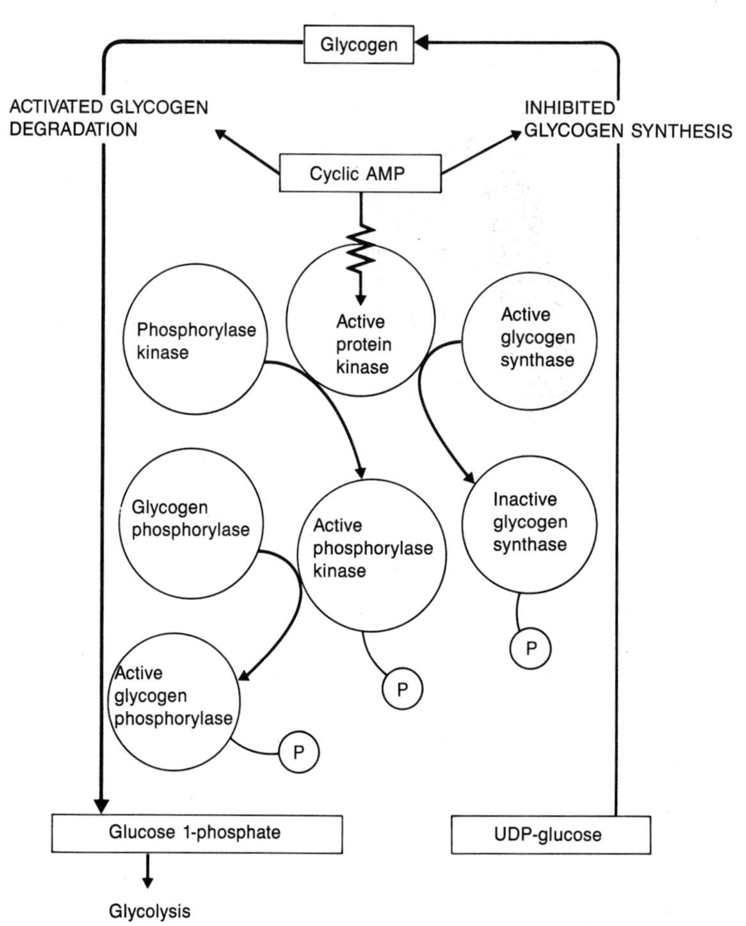

Fig. 4.12. Regulation of blood sugar level. How an increase in the level of cyclic AMP in skeletal muscle (induced by binding of adrenaline to cell-surface receptors) stimulates glycogen degradation and inhibits glycogen synthesis. The binding of cyclic AMP to a specific protein kinase activates this enzyme and causes it to phosphorylate and thereby activate the phosphorylase kinase. The latter, in turn, phosphorylates glycogen phosphorylase, the enzyme responsible for the degradation of glycogen, and thereby activates it. The same cAMP-dependent protein kinase phosphorylates and thereby *inactivates* glycogen synthase, the enzyme responsible for glycogen synthesis.

first person to demonstrate experimentally the electrical nature of lightning. During a thunder storm, he stood at the foot of a hill called the Hainberg in Göttingen, flying a kite attached to thin metal wires – by no means a safe experiment! – and was able to measure the amount of current conducted through the wires. He also had some important ideas about protection from lightning: "If I lived in a country where the storms raged as they do in, say, Carolina, I would construct a cage out of golden

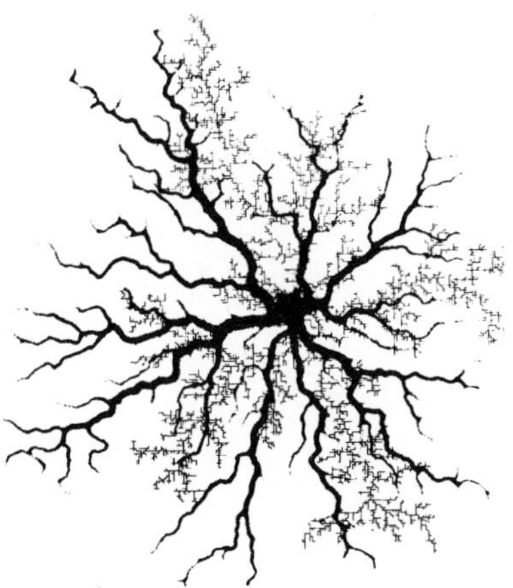

Fig. 4.13. Why doesn't lightning move in a straight line? Left: A detail from Hogarth's fourth plate of "Der Weg des Liederlichen" [The Path of the Loose Woman] in a copy engraving by E. Ripenhausen. Lichtenberg's comment: "[The lightning] is uncertain where it wants to go, after all. . .the zigzag is really the line of indecisiveness, and I am therefore somewhat reluctant to disapprove of that good woman, who believes that lightning zigzags and changes its direction so often because it always turns away from places where people would still have mended their ways in a hurry."
Right: Lichtenberg discharge figure on a glass surface in sulfur hexafluoride gas. The photo should be compared to a (grid) simulation, which takes into account that fluctuations also play a role in a dynamical process. They lead to branching and are governed by probability distributions. The similarity is striking. The figures turn out to be fractal structures (theory of Benoit Mandelbrot).

bars above my house [*the Faraday cage!*]. We know only two propositions with certainty about the nature of lightning: 1. It is electrical. 2. When it encounters a continuous stretch of metal, it follows this and, if the metal extends into the ground, it enters there as well. . .it is completely sufficient if only the chimney, the ridges, and all projecting edges of the building are covered with overlapping strips of lead or copper and grounded. The high and pointed rods can be completely omitted."[17]

Concerning lightning, he adds: "Is it not highly unusual that lightning, which moves with such rapidity, never or seldom goes in a straight line but allows itself to be easily deflected? One readily sees, then, that the bolt does not go very far into the distance, but instead from one point to another in close proximity."

The nature of lightning is clearly recognized here. Electrical charge jumps from one ionized particle to another. "The bolt does not go very far into the distance." His humorous commentary on the Hogarth copperplate engraving is given in the caption to Figure 4.13. Also famous are the Lichtenberg figures (Fig. 4.14). They are the ionization traces left by electrical charges passing through a nonconducting plate.

Fig. 4.14. Lichtenberg figure. A plate of Plexiglass (13×9×2 cm) was irradiated perpendicular to the plane of the picture with high-energy electrons (2.8 MeV), which penetrate to a depth of 7 mm. At first they remain trapped, because Plexiglass is a good insulator; subsequently, however, they manage to penetrate to a defect zone caused by grain formation at the edge of the plate (right in the picture). The Plexiglass melts and vaporizes along the discharge path, resulting in a pattern of fine branching.

Fig. 4.15. Lightning above Lugano.

Atmospheric lightning is shown in Figure 4.15. Lightning "evolves"; that is, it springs from one point to the next. At each point, an irreversible decision is made as to the subsequent path to be taken. This decision is understandable only in statistical or quantum-mechanical terms. Whenever matter and energy are simultaneously

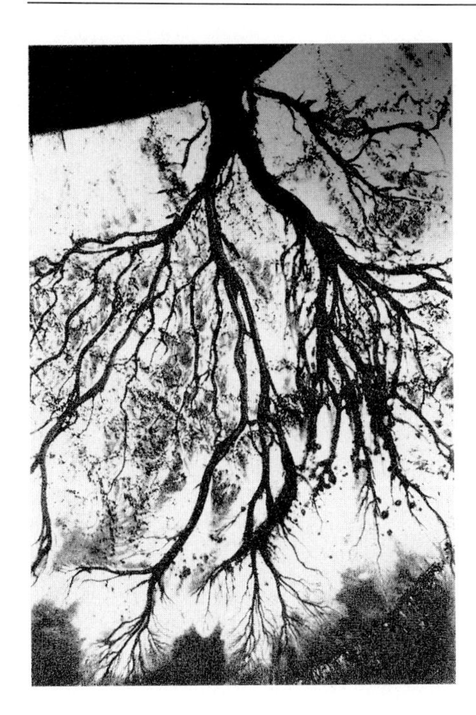

Fig. 4.16. Delta of the Colorado River at the Gulf of California.

transported through a medium in a system far from equilibrium, that is, at a high energy level, such "lightning" occurs. The time scales of these processes can be totally different.

The delta of the Colorado River where it flows into the Gulf of California is shown in Figure 4.16. This, too, represents a "decision tree." The individual branches of the delta "decide" which way they will flow under the influence of the current (the energy-supplying component), the particles of sand that are swept along, the sandbanks, the wind, and the tides. It is impossible to predict which direction each arm will take. Given a knowledge of the nature of the soil (mud, moor, etc.), currents, climatic factors, and average water temperatures, as well as empirical data on the formation and shape of deltas, a geophysicist could determine whether the delta would be swampy or consist of many arms or have just a few main channels, whether vegetation could grow there and reinforce the river banks, or whether the delta would form anew each year. However, such scientific predictions never hold for the individual events and forms in a river delta, because it, too, is a dissipative dynamical structure as described by Prigogine (see chapters 4 and 9).

The same holds for a tree, as shown in Figure 4.17. It is true that the basic shape of a tree is dictated by a genetically fixed program – a pine is always different from a poplar, which, in turn, is different from a beech tree. Within the range of variation allowed by this genetic system, however, the form of the tree cannot be predicted exactly. When and where the bud for a new branch will emerge, how fast it will grow,

Fig. 4.17. A tree standing alone.

to what extent it will grow at the expense of other branches and block the light from reaching them, the effect of its habitat, the influence of climatic factors and the seasons, none of these is predictable a priori. A tree, unfolding according to a certain genetic program, is also a system with branch points; materials are transported at a high energy level (that of life), energy is dissipated, and irreversible decisions are made. In principle, a tree could be regarded as a bolt of lightning moving extremely slowly. The time scale is about 10^{12} times slower.

Structure and Fluctuations – Prigogine's Theorem

Since the beginning of this century, since Boltzmann's work (see chapter 8), we have known that we live in a nonequilibrium world. The basic structures of the cosmos evolve. As just mentioned, the flows of energy in evolving "tree systems" are associated with physical and chemical events, so something new is always being formed. If the energy flow ceases, thermodynamic equilibrium is immediately established; the system is dead. If it were possible for the substances in my body to attain thermal equilibrium in a few seconds, for example, I would vanish in a puff of smoke and vapor, leaving behind nothing but a heap of ashes.

Everything we do in science requires simplification. In the 19th century, when the steam engine was invented and analyzed, it was reasonable to postulate reversible thermodynamics and the conservation of energy (first law) and to regard irreversible systems as a special case. At the end of the 20th century, the great scientific themes confronting us include life, evolution, the intricate processes in the brain, and the formation of the cosmos. It is reasonable, then, to regard classical thermodynamics as the exception and the energetics of irreversible systems, where the second law applies, as the rule.

Life arose. It was not there before and how it originated is not yet fully clear to us. Still, we have learned something about Eigen's theory of hypercycles. We have also discussed Darwin's ideas and the phylogenetic trees depicting the origin of individual species. Many complex processes show such "tree structures." Indeed, complex evolving processes in the form of tree structures represent a general principle. What are the implications of this realization?

We are accustomed to thinking in terms of simple Newtonian trajectories. Furthermore, our mental pathways have been subjected for decades to a reductionist pedagogy; they follow parabolic trajectories, like tossed stones. The initial speed and direction determine once and for all how far the stone flies and where it lands.

The acceleration and the braking distance of our car are calculable quantities, everyday facts of life. Newtonian processes are continuous events and can be reproduced at any time and at any place. But this is not true of the events in highly complex systems such as life or, equally, high-energy physics, systems involving turbulence, or elementary particle physics. Nonetheless, life is something too close and self-evident and the systems mentioned above are far from being directly observable. Therefore, our spatial-temporal viewpoint does not seem to require much modification. Systems with tree structures have, as already pointed out, branch points, where a choice is made among alternative, equally probable paths. Which path is taken cannot be predicted. Strictly deterministic initial conditions, even when all parameters are specified, do not allow predictions to be made about the branch points. The result is indeterminism. These branch points are called fulguration points (from the Latin *fulgur*: lightning); they bear an analogy to lightning.

In the previous section, we learned something about lightning. All of the complex systems discussed here, no matter how different their constituents, are in principle identical. They are formed by nonreproducible processes; they unfold, live, grow old, and die. Why? In linear systems every process is repeatable and reversible. Time in the classical mechanics of linear systems is reversible; its structure is not polar. Newtonian systems do not grow old.

By contrast, it is not possible a priori to go backwards in a tree system with fulguration points. An irreversible decision has been made at the fulguration point. The time axis in a tree system is irreversible; in such systems, time takes on a entirely new meaning. Or has it simply reacquired its old meaning? We will come back to this point in chapter 8.

The fulguration points have important consequences for the predictability of events. Prigogine writes[18]: "The basis of the vision of classical physics was the conviction that the future is determined by the present, and therefore a careful study of the present permits the unveiling of the future. At no time, however, was this more than a theoretical possibility. Yet in some sense this unlimited predictability was an essential element of the scientific picture of the physical world. We may perhaps even call it the founding myth of classical science."

Prigogine developed the theory of dissipative structures far from equilibrium and, in his Nobel Prize lecture, has the following to say[19]: "Far from equilibrium, there appears an unexpected relation between chemical kinetics and the 'space-time structure' of reacting systems. It is true that the interactions which determine the values of the relevant kinetic constants and transport coefficients result from short-range interactions (valency forces, hydrogen bonds, van der Waals forces). However, the solutions of the kinetic equations depend, in addition, on global characteristics."

In Lichtenberg's analysis of lightning – "the bolt does not go very far into the distance, but instead from one point to another in close proximity" – this notion is already clearly expressed. The "short-range interactions" are the fluctuations of the electrically charged molecules of air, the "global characteristics" is the electric field between the thunder clouds.

Prigogine remarks further: "This dependence, which on the thermodynamic branch, near equilibrium, is rather trivial, becomes decisive in chemical systems under conditions far from equilibrium. For example, the occurrence of dissipative structures generally requires that the system's size exceed some critical value. The critical size is a complex function of the parameters describing the reaction-diffusion processes. Therefore, we may say that chemical instabilities involve long-range order through which the system acts as a whole."

"There are three aspects that are always linked in dissipative structures: the function as expressed by the chemical equations; the space-time structure, which results from the instabilities; and the fluctuations, which trigger the instabilities. The interplay between these three aspects leads to most unexpected phenomena, including 'order through fluctuations...'"

$$
\begin{array}{ccc}
\text{Function} & \rightleftharpoons & \text{Structure} \\
\searrow\!\!\nwarrow & & \swarrow\!\!\nearrow \\
& \text{Fluctuations} &
\end{array}
$$

In general, we obtain successive bifurcations when we increase the value of a characteristic parameter. In Figure 4.18, we find a single solution for the value λ_1, but many solutions for the value λ_2.

Prigogine notes: "It is interesting that bifurcation introduces, in a sense, 'history' into physics. Suppose that observation shows us that the system whose bifurcation diagram is represented [by Fig. 4.18]...is in the state C and came there through an

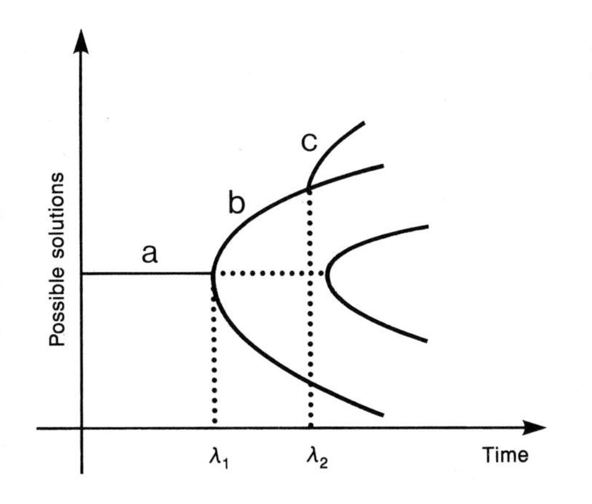

Fig. 4.18. Successive bifurcations in an evolving system.

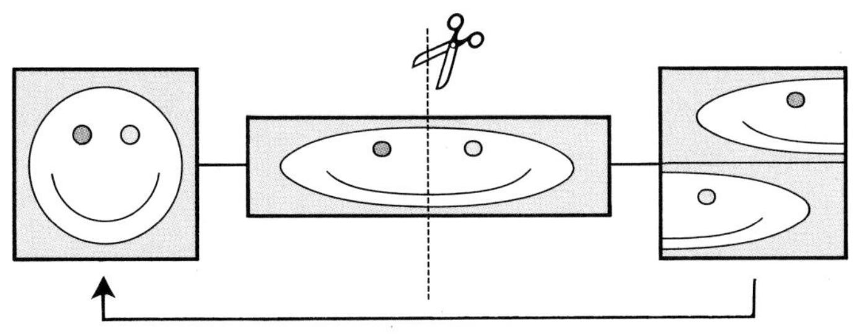

Fig. 4.19. Baker transformation of a square design. The square is stretched to twice its length and half its height and then cut through the middle. The two parts are recombined to form a square. As the process is reiterated, the position of a specific point is followed. In our example, an "unusual" discontinuity appears at the 10th transformation, a bifurcation at transformation 16, 27, 30, 31, and 37.

increase in the value of λ. The interpretation of this state c implies the knowledge of the prior history of the system, which had to go through the bifurcation points A and B. In this way we introduce in physics and chemistry a historical element, which until now seemed to be reserved exclusively for sciences dealing with biological, social, and cultural phenomena."

"Every description of a system which has bifurcations will imply both deterministic and probabilistic elements. As we shall see in more detail [chapter 5], . . .the system obeys deterministic laws, such as the laws of chemical kinetics, between two bifurcation points, whereas in the neighborhood of the bifurcation points fluctuations play an essential role and determine the branch the system will follow."

How does one describe the trajectories of branched systems in which "chaotic points" or fulgarations appear? Normal coordinate transformations are continuous. A tossed ball does not follow its trajectory in a zigzag fashion. The mathematical description of branched systems requires a trajectory with discontinuities; Prigogine

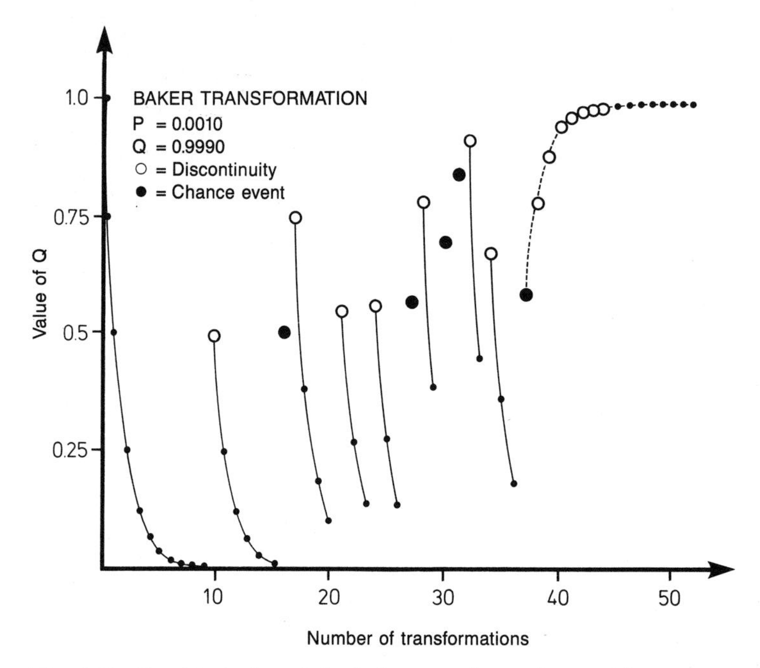

Fig. 4.20. Graphical plot of the baker transformation of Figure 4.19.

suggests the so-called baker transformation as a discontinuous transformation (Figs. 4.19 and 4.20). It consists in making a deterministic system indeterministic by simple geometrical operations: a design (here, a face) is stretched like pasta dough, cut through the middle, and then put back together to form a square. This is a simple, strictly deterministic procedure. If a single point – say, the pupil of the eye – is followed in a coordinate system, then unpredictable discontinuities are found. The point begins to spring around and finally vanishes from the system.

This is one of many systems, called Bernoulli systems, in which mathematical operations have an unpredictable outcome[20,21,22]. We will examine such systems in greater detail in chapters 5 and 6.

We are accustomed to having our daily world, our "little village," function in a linear way without discontinuities or sudden jumps. We learn to expect processes to function, to be continuous, and our technical world daily reinforces this attitude. It is a magnificent achievement of the sciences to have found the laws and rules within nature that make predictions possible. But at crucial points – wherever something new emerges – everything is suddenly open and undecided. Despite "the weighing of all possibilities and the consideration of all physical relations" in a system, an unpredictable decision is made.

Emily Dickinson[23]

The farthest Thunder that I heard
Was nearer than the Sky
And rumbles still, though torrid Noons
Have lain their missiles by –
The Lightning that preceded it
Struck no one but myself –
But I would not exchange the Bolt
For all the rest of Life –
Indebtedness to Oxygen
The Happy may repay,
But not the obligation
To Electricity –
It founds the Homes and decks the Days
And every clamor bright
Is but the gleam concomitant
Of that waylaying Light –
The Thought is quiet as a Flake –
A Crash without a Sound,
How Life's reverberation
It's Explanation found –

5. Mathematical and Physical Models of Deterministic Chaos

A dialogue between Georg Christoph Lichtenberg and his colleague Ludwig Wittgenstein on logic, chaos, and the properties of living systems. Both men were *outsiders*, particularly *on their academic faculties. Whereas the other faculty members were on congenial terms with one another, extended mutual invitations, served as godfathers, and sent one another sausage soup after a slaughter, these two men were always involved in petty quarrels, cast a critical eye on each other's work, and sought errors in each other's books.*

WITTGENSTEIN[1]: *I think that I have summarized my view of philosophy by stating that philosophy really ought to be expressed in verse only. It seems to me that this should reveal to what extent my thinking belongs to the present, the future, or the past. For I have also admitted to being a person who cannot quite do what he wishes he could do.* What do you have to say about this, Herr Colleague?

LICHTENBERG: *Of course, I, too, have learned to play a few pieces on the metaphysical piano,* my dear Herr Wittgenstein, but our mathematical logic is so much more lucid. *It is conceivable that many ideas of Kantian philosophy have not really been understood by anyone. Yet each person thinks that others understand them better than he does and is therefore content with a vague idea, or even believes it is his own inability that prevents him from comprehending as clearly as others do.*

WITTGENSTEIN: Inferiority complexes have no place in this matter, my dear man. *If a fallacy is expressed boldly and clearly but once, quite a lot is thereby accomplished.*

Only when one thinks more wildly than philosophers is one able to solve their problems. If people did not do silly things now and then, nothing intelligent would ever happen.

LICHTENBERG: I absolutely agree, Herr Colleague. *Just as Linnaeus did for the animal kingdom, one could set up a class in the realm of ideas and call it chaos. It is not so much the great ideas of general profundity that belong there – stardust with sun-dappled spaces of the vast whole – but rather the small infusing ideas that latch on to everything with their tail and often live in the seeds of great ideas. Every man, sitting still, sees a million such ideas coursing through his head.*

WITTGENSTEIN: Precisely, Colleague Lichtenberg: *To philosophize, it is necessary to descend into the old chaos and to feel at home there. Always descend from the barren heights of genius into the greening valleys of stupidity.*

LICHTENBERG: Your unconventional comments fascinate me, although – just between us – you are quite frivolous for a scientist. *It would be wonderful if one could learn to discover by following certain rules. For instance, what the hottest topics are or how to induce reason; this would be just as much of a discovery as how to increase the size of animals or how to make shrubs as large as oak trees. It seems as if a kind of chance underlies all discoveries, even those apparently made with much effort. Likewise, it seems to me that the improvements attainable by countries through rational considerations are just minor changes; we make new species, but we cannot create genera, which are the product of chance only. Experiments should therefore be undertaken in science and the grand events must be given due time . . . Something I said elsewhere applies here as well: one should not say "I think," but rather "It thinks," just as one says "It lightnings."* But let's talk a little about my dear mathematics, Herr Colleague.

WITTGENSTEIN: Hmm, do you know what, Lichtenberg? *I can climb the peak of mathematics only slowly with a full philosophical knapsack.*

LICHTENBERG: You aren't usually so fainthearted, reverend sir. *The great progress made with mathematics is due alone to its independence from everything that is not mere quantity. That is, everything besides quantity is completely alien to it. Since it concerns itself with quantity alone and requires no outside help, but is solely a development of the laws of the human mind, it is not only the most certain and reliable of all human sciences, but also without a doubt the simplest. Everything needed to pursue it further is in man himself. Nature equips every intelligent person with the complete apparatus; we receive it as a dowry. In this very way it becomes the simplest of all sciences, in that no other allows us to proceed so far or to hope for so much.*

WITTGENSTEIN: *In no religious doctrine have so many sins been committed through the misuse of metaphysical expressions as in mathematics.*

LICHTENBERG: That doesn't really say much, since *mathematical order is the basis of all things. During my illness in January and February of 1790 I often gazed at the canopy of my bedstead, which was covered with a minute floral pattern. Each tiny flower lay at the point shared by two lines intersecting at about 60°. This resulted in a large number of rhombuses. Whenever I focused on only one rhombus of about one square inch, or 4, or 9 square inches, and so forth, the whole surface appeared at once to consist of rhombuses, all of the same size as the first. This also occurred if, instead of rhombuses, I tried rhomboids. These were thus patterns made up of objective and subjective elements at the same time. When I tried a new one, it was always somewhat difficult at first, but, once begun, the whole seemed to crystallize spontaneously all at once. I think this phenomenon could be applied to higher things. In a large number of equally spaced points I might be able to visualize all kinds of signs and patterns, which, once perceived at one end of the surface, would soon be found everywhere else. In this way, order could be discerned in the greatest disorder, just as we perceive images in the clouds and in bright-colored stones.*

WITTGENSTEIN: *The mathematician Pascal, who was fascinated by the beauty of a theorem of number theory, was equally fascinated by natural beauty. It is wondrous, he said, to see what sublime properties numbers have. It is as if he were fascinated by the regularities of a kind of crystal.*

LICHTENBERG (satirically): *The noble simplicity seen in the works of nature is due all too often to the noble nearsightedness of the observer.*

WITTGENSTEIN: Oh no, Colleague Lichtenberg, there you are totally wrong: *There are simple principles of nature. But life, especially human life, is* fundamentally *complex. Life is like a path along a mountain ridge; to the right and left lie slippery slopes, down which you slide, in one direction or the other, without stopping. Again and again, I see people slipping like this and say "How can a person help it!" And that means "to deny free will."*

LICHTENBERG: You see, we agree after all: *Man is a masterpiece of creation if only because, in the face of all determinism, he believes that he acts as a free being.*

WITTGENSTEIN (laughing): Tut, tut, you are being a sophist, Lichtenberg. *When one cannot untangle a knot, the most clever thing to do is to recognize that fact and the most honorable thing is to acknowledge it.*

LICHTENBERG: Indeed, I do, I do, Herr Wittgenstein. *It is not nearly as bad to explain a phenomenon with a little bit of mechanics and a strong dose of the incomprehensible as to try to explain it by mechanics alone. That is, the docta ignorantia is less cause for shame than the indocta. All motions in the world have their origin in something without motion, so why shouldn't the universal force be the cause of my thoughts as well, just as it is the cause of fermentation?*

If, someday, a higher being tells us how the world originated, I would really like to know whether we would be able to understand. I don't think so. Origin could hardly give rise to anything, since this is mere anthropomorphism. It may even be that, except for our mind, nothing exists that corresponds to our notion of origin.

WITTGENSTEIN: *How can one speak of "understanding" or "not understanding" a proposition; doesn't it become a proposition only after one understands it?*

LICHTENBERG: True. In fact, I wanted to say that the origin of life is an unanswerable question, because life is a system; *the creatures do not make up a chain, as the poets (Pope) often express themselves, but rather a net, for they often come back together from the side as well. This is clearly shown by the transitions of animals and stones from one species to the next and from one genus to the next.*

WITTGENSTEIN: Correct, Herr Colleague, the whole is more than the sum of its parts. *Raisins may be the best part of a cookie, but a sack of raisins is not better than a cookie; and whoever is in a position to give us a sack of raisins is not necessar-*

ily able to bake a cookie, let alone something better. A cookie is not the same thing as diluted raisins.

LICHTENBERG: I think we understand each other now, my dear Herr Wittgenstein. But let's return to the starting point of our discussion, the role of philosophy in the solution of the world's mysteries.

WITTGENSTEIN: *There is no mystery. If a question can be posed at all, it can be answered. For doubt exists only where a question exists, a question only where an answer exists, and this only where something can be said.*

We sense that, even if all scientific questions were to be answered, the problems in our life would remain completely untouched. Of course, no question would remain then and that, in fact, is the answer. The solution to the problem of life is revealed when this problem vanishes. However, some things cannot be said. This shows that a mystical element is involved.

The correct method of philosophy would be the following: To say nothing other than what can be said, namely, propositions of natural science – that is, something that has nothing to do with philosophy – and then, whenever somebody wants to say something metaphysical, to demonstrate to him that he has not given a meaning to certain signs in his proposition. This method would be dissatisfying to the other person – he would not have the feeling that we were teaching him philosophy – but it would be the solely rigorous approach. My propositions offer an explanation insofar as he who understands me recognizes them as nonsense in the end, when, through them, he has passed beyond them. He has to surmount these propositions to see the world truly.

What cannot be said should be left silent.

What Is Chaos? Bifurcation Points of Dissipative Structures

The word chaos has become rather worn-out through common usage. First of all, then, let us trace its original meaning. The word, derived from the Greek, originally meant something abysmal, gaping – the emptiness of space. In ancient cosmogonies – as early as the pre-Socratics, but also in the much older biblical story of creation – this barrenness and void is the fundament of all becoming, the ultimate origin of the cosmos. Chaos and cosmos, formless being and ordered structures, are thus closely related. This interpretation of chaos is still common in more modern philosophies. Schelling regarded chaos as the "metaphysical unit of potentialities." The modern sciences of dynamical processes – through which something new arises –

have adapted this old concept of chaos to their own use[2,5]. In the meantime, colloquial usage has worn down the concept of chaos to mean simply the undesired breakdown or decay of order (traffic chaos, chaotic discussions, chaotic people, etc.).

Of course, chaos can arise through the decay of order. In many dynamical processes, as will be shown later, phase transitions pass through chaotic situations, which then stabilize themselves in a new, higher order. This is what occurs in evolving systems at all branch points or bifurcations (from Lat. *furca*: forked, actually two-pronged), as we have already seen in chapter 4. Chaos and order, then, are not just a pair of concepts; they have a dialectical or functional relationship to each other.

The expression deterministic has a clearer definition. Deterministic means predetermined and predeterminable. In a positivistic worldview of physics – which, as we have seen, is outdated – it was believed that all parameters of an object, a motion, or a living thing could be determined so precisely and completely that its future, no matter how complex, could be predicted by means of differential equations – for example, the course of a moving body, a flying projectile, an amoeba, a bolt of lightning. Such courses of motion are called trajectories; they might equally be referred to as pathways of development. However, trajectories follow deterministic paths only in those systems where linear differential equations (equations in which differential quotients appear as variables) are applicable, often only approximately. In nonlinear systems, they pass through one or more bifurcation points and thereby become indeterministic. We have already seen qualitative examples of this in the preceding chapter. As early as 1892, the great French mathematician Poincaré worked out the mathematical groundwork for the treatment of nonlinear systems[3]. But only in 1963 did Poincaré's ideas find practical application; the American meteorologist E. N. Lorenz used them to create mathematical models for the calculation of weather patterns[4]. In these models, Lorenz simulated the most important parameters of meteorological situations as well as how these parameters interact. He found that a set of even three nonlinear, coupled, first-order differential equations led to totally chaotic trajectories. Thus, deterministic chaos means that the use of deterministic equations of motion leads nonetheless to a chaotic trajectory. Weather literally thwarts our predictions.

Several systems of this kind will be discussed in this chapter. It is fundamentally impossible to predict when a potentially chaotic system will become chaotic. Indeed, this unpredictability is a salient feature of its behavior. It is much easier to list what does not lead to chaos[5].

Systems described by linear differential equations are solvable arithmetically. In particular, if several linear differential equations are required to describe a system, they can be solved by a mathematical method called Fourier transformation. Such equations do not lead to chaos, however. Nor is chaos due to external influences or to situations in which the number of variables (degrees of freedom) needed to describe the system is simply too large. These factors may very well set practical limits to the description of such systems, but not fundamental ones. Nor does chaos arise from the uncertainty underlying the statistical character of quantum mechanics.

Potentially chaotic structures are always nonlinear structures that involve feedback coupling and are strongly dependent on initial conditions. The global structure arising during the process in question is influenced in an unpredictable way by the fine details of the initial situation. Lorenz described this as a butterfly effect: a single flutter of a butterfly's wings can (but, of course, need not necessarily) result in a complete change in global weather patterns.

Morphogenesis and Catastrophe Theory

In principle, there are four distinct ways that a moving body can behave in a system with potentially nonlinear structures (cf. Fig. 5.1). First, it can move chaotically. Second, it can move toward a center. Third, it can undergo simple oscillation. Fourth, it can oscillate with a higher period.

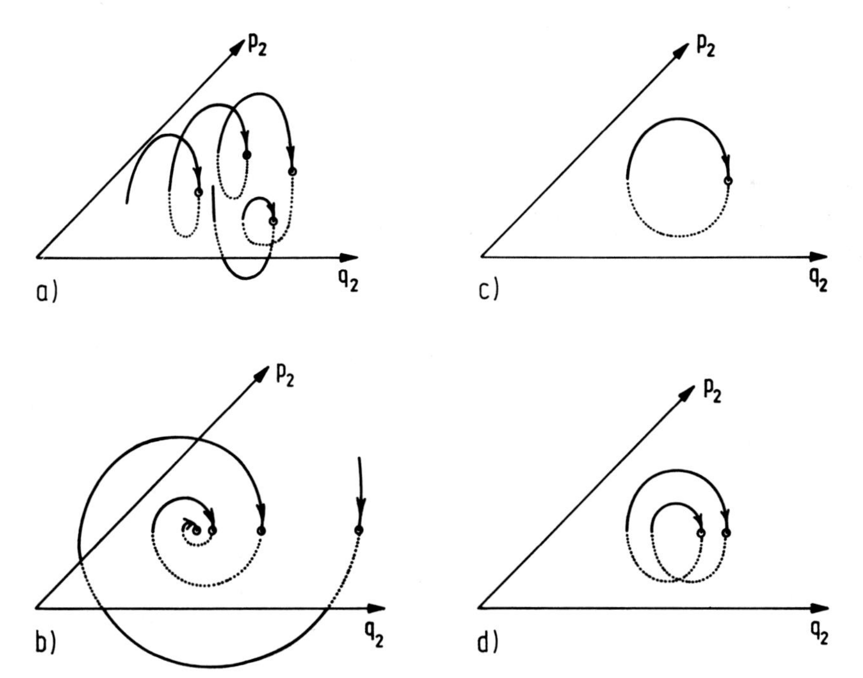

Fig. 5.1. Qualitatively distinct trajectories in Poincaré plots. (a) Chaotic motion. (b) Motion toward a center, approached asymptotically. (c) Periodic motion or limit cycle. (d) Higher-period limit cycle.

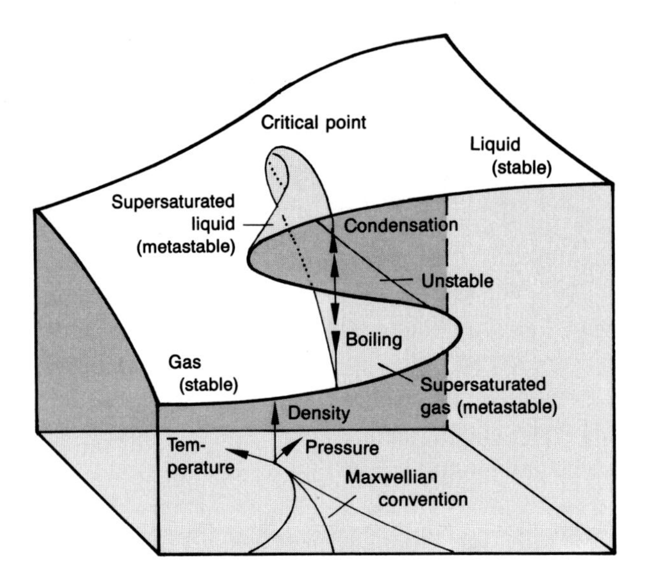

Fig. 5.2. Phase transitions between liquid and vapor states in a cusp catastrophe.

The mathematician René Thom, the creator of catastrophe theory, has attempted to trace the generation of forms in nature to symmetry breaking and "branching catastrophes" in order to provide a mathematical description of morphogenesis, which was discussed in chapter 1 (see also chapter 7). The term "catastrophe," now well-established in mathematics, exaggerates the situation somewhat. What is actually involved is the mathematical description of fundamental discontinuities, points at which a trajectory or a system of trajectories abruptly changes or branches. The simplest discontinuity is the branching or bifurcation (forking) occurring in any tree structure. There, it is referred to as a fold catastrophe and is describable in two dimensions. The catastrophe of the next higher dimension, three dimensions, is the cusp catastrophe, in which a system can move in various directions from a cusp (cf. Figs. 5.2 and 5.3).[6] Many systems with discontinuities are describable in this way. Examples include the sudden boiling of superheated water and stock market crashes.

In his theory, Thom discusses far-reaching applications such as embryogenesis, cell division, dreaming, play, the development of human speech, the structure of human society, and more. Though still rather general, the discussion is intellectually stimulating to follow. In fact, there have been several attempts to describe concrete systems. Figure 5.2 shows the phase behavior of a system between the liquid and vapor states according to the cusp catastrophe model. The controlling factors here are temperature and pressure. Normally – that is, outside the cusp region – boiling and condensation occur at the same temperature – for example, 100 degrees Celsius for water. Under certain conditions, however, the vapor can be cooled below its dew point or the liquid heated above its boiling point (superheating). The surface describing this kind of state extends into the fold region or even to the cusp itself. The result

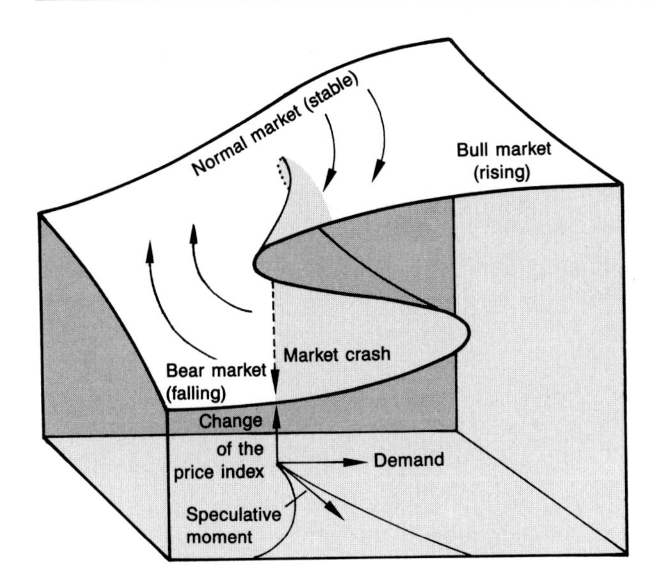

Fig. 5.3. "Phase space" of the stock market under the influence of demand, change of the price index, and speculation. A jump from the upper curve to the lower corresponds to a stock market crash.

is hysteresis (retardation) and eventually even a "catastrophic" transition, namely, violent boiling. The model may also be applied to systems in society, such as the stock market. The controlling factors here are the demand for stocks and the speculative strategies of the seller. In Figure 5.3, for example, strong speculation could cause the system to reach the surface above the fold or even the cusp itself, leading abruptly to a stock market slump or even a stock market crash.

The Three-Body Problem – The Double Pendulum

Poincaré's interest in the problem of "branching systems" was motivated in part by the so-called three-body problem, long important in the mechanics of heavenly bodies. Not even today can one describe the motions of three mutually interacting bodies – say, the sun, the earth, and Mars – in such a way that, at any point in time, a deterministic prediction can be made about the fate of the three bodies at some future time. This problem arose shortly after classical Newtonian mechanics had celebrated one of its greatest successes, the discovery in 1846 of the planet Neptune, whose existence was predicted on the basis of small deviations in the orbit of Uranus.

At that time, the Swedish Academy of Sciences posed the prize question: How stable is our solar system? Poincaré was able to show that a deterministic answer to this question is fundamentally impossible. In a certain sense, then, the answer was nega-

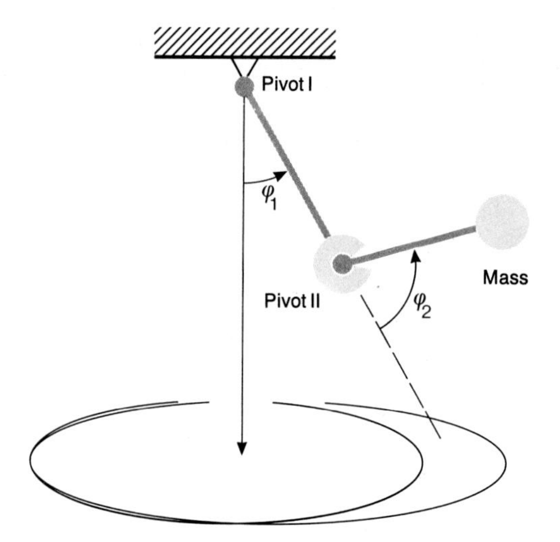

Fig. 5.4. The double pendulum.

tive. The positive consequence, however, was the creation of bifurcation mathematics and the discovery of deterministic chaos. Examined more closely, the "Swedish Academy question" about the stability of the planetary system implies a more far-reaching question about the evolution of this system. Its answer requires a completely new intellectual concept – Poincaré's bifurcation mathematics. Similarly, an answer to the "Göttingen Academy question" (see chapter 2) requires the new concepts of feedback amplification, the hypercycle (cf. reference 7 in chapter 4), and feedback catastrophe (cf. references 14 and 19 in chapter 8).

The pendulum offers a convenient experimental device to study the three-body problem. A normal pendulum on earth represents a two-body problem governed by the mutual attraction between the pendulum's mass and the earth. The mass of the pendulum is suspended from a framework and thereby prevented from plunging toward the center of the earth. Instead of falling straight downward, therefore, it oscillates harmonically around the pendulum axis (see, for example, Fig. 5.1c). The two bodies, then, are the mass of the pendulum and the earth. Similarly, the three-body problem may be studied by using a double pendulum, a pendulum to which a second pendulum is attached. Alternatively, a double pendulum (Fig. 5.4) could be described as a pendulum with a knee joint. The two pendulums are free to swing independently of each other, but are nonetheless coupled. This very complicated behavior is best understood in terms of the representation given by Poincaré; the motions are not recorded continuously, but only at certain moments that are somehow characteristic of the system – for example, when the two arms are fully extended. In this case, one plots the angle and the corresponding momentum. In principle, the following situations are conceivable:

1. Nothing moves. This is usually the case when the system finds itself in a stable position. That is, it has run down and no longer receives any impulse or "kick."

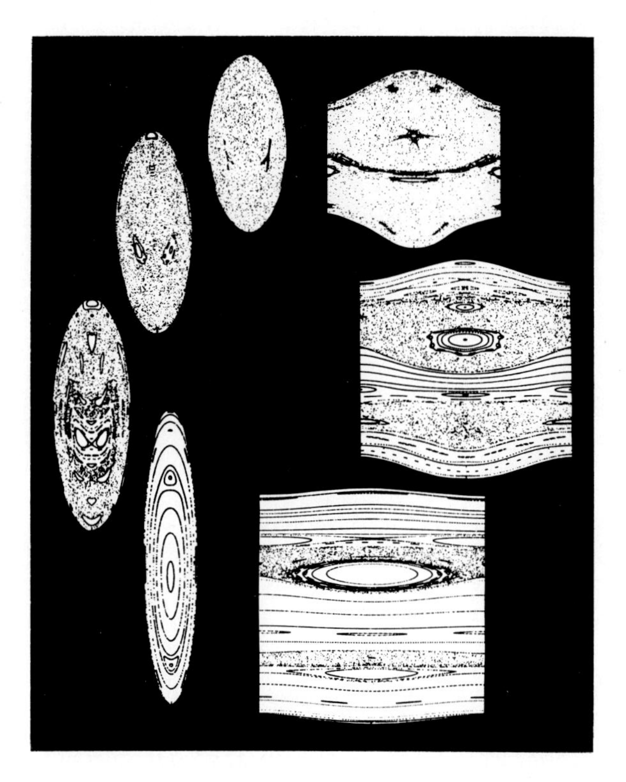

Fig. 5.5. "Orbits" of the double pendulum in a Poincaré plot. Bottom left: Oscillations at lower energy in defined orbits. With increasing energy (clockwise) the points leave the orbits and start to cover the whole surface of events in a chaotic manner[7].

There is another conceivable position, however, in which one or both arms of the pendulum extend straight upward. This state is unstable and very difficult to verify, since even the slightest jar would cause the arms to fall.
2. The double pendulum oscillates periodically. This situation is encountered when the oscillations of the two parts of the pendulum are in a constant ratio to each other – for example, the (shorter) lower pendulum oscillates twice as fast as the upper.
3. A third possibility is quasi-periodic motion. That is, the oscillation ratio is constant but irrational. This situation can lead to periodic motion as well. The corresponding forms of motion are traced in a Poincaré map in Figure 5.5. The elongated ellipsis at the bottom left shows the system at lower energy. The lines are continuous, even though they are built up over long periods through the accumulation of many points. This shows that the system passes repeatedly through the same states; at low energy the double pendulum has certain "orbits."
4. At higher energy – that is, angular momentum – the pendulum starts to show chaotic behavior. If given a strong enough impulse, the double pendulum passes through situations that lie outside the defined elliptical orbits. Although there are isolated islands of order (in the second and third ellipse viewed clockwise), the whole surface of events gradually fills with points. The system becomes chaotic. Finally, at a certain energy somewhere in between, the regions of order vanish. At still higher

energies (top right and just below), however, order emerges again. The islands of order become larger and larger until only a few bands of chaos remain. The explanation is simple. For high rotational momentum, the third body, the earth with its gravitation, plays an ever-smaller role. That is, the rotational momentum and the centrifugal force of the double pendulum become so large that the earth's gravitational attraction is negligible in comparison. The system is reduced by one dimension and approximates a two-body problem.

The Rings of Saturn

Anyone who has ever gazed through a telescope at Saturn never forgets the sight of a wondrous, shimmering reddish sphere surrounded by strange-looking rings resembling a hallow.

The rings of Saturn are made up of dust particles, grains, and small chunks of matter, which orbit the giant planet individually like mini-moons and, from far away, resemble a disk. Unlike our own moon, they have not aggregated to form a single compact body, although Saturn does have several true moons. When the disk is viewed at greater telescopic magnification, gaps become apparent at certain intervals. Some regions, particularly the so-called Cassini division (see Fig. 5.6), appear to have been swept clean of all matter.

It is now known that the Cassini division, as well as the other large gaps in the ring, correspond to "resonance zones" of Mimas, one of Saturn's moons. The parti-

Fig. 5.6. Saturn's ring with the Cassini division.

cles missing from the Cassini division would have orbited with exactly half the period of Mimas or, so to speak, one octave higher. Such behavior is called resonance, not only in acoustics but for periodic processes in general. Further gaps correspond to resonances of higher order. What is the explanation of these observations?

There are other intriguing numerical ratios in our planetary system. For example, Jupiter and Saturn orbit the sun with periods whose ratio is exactly 2:5. The three-body system Sun-Jupiter-Saturn thus has a "time solution" with the ratio Jupiter : Saturn = 2:5. Over billions of years the orbital periods settled into what is apparently the most stable ratio.

Heinz O. Peitgen and Peter H. Richter, the source of many of the ideas discussed here, have developed a simple method of calculation to simulate complex dynamical systems in such a way that every point is related to the following point[7]. In other words, they have formulated rules specifying how each point arises from the preceding one. The structures thereby obtained are similar to those found for the double pendulum. This mapping approach thus allows one to follow the course of dynamical processes on the display screen of a computer. After many iterations, it becomes readily apparent whether a sequence of points is either regular or chaotic. The following discussion closely follows the presentation of Peter Richter[7].

In order to understand the "planetary resonances," the dynamics of a pendulum will be explained once again. In Figure 5.7 the dynamics of a spring and a pendulum are compared. Because of the proportionality between the applied force and the vibrational amplitude (extension) of the spring, it exhibits very simple dynamics. On the other hand, the pendulum, owing to the cosine function occurring in its potential, gives a much more complex map. Let us examine these maps more closely.

First of all, we notice that, near the center, the oscillations of the spring are similar to those of the pendulum. For displacements by small angles, in fact, the oscillation

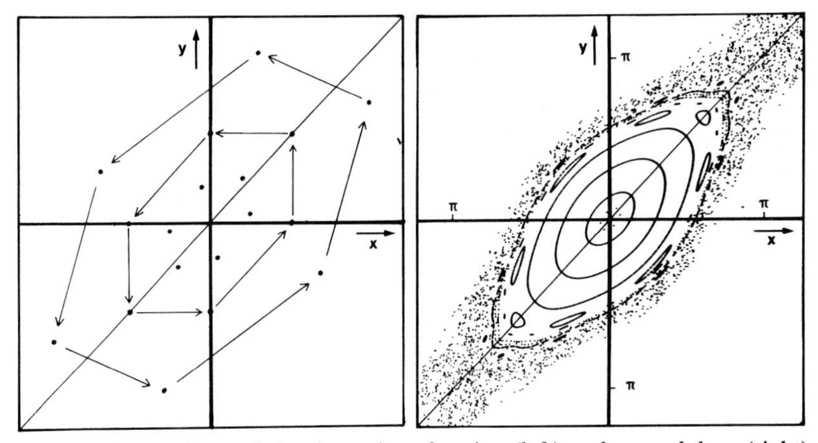

Fig. 5.7. Comparison of the dynamics of spring (left) and a pendulum (right).

of a pendulum is exactly like that of a spring. By convention, the point $x = y = 0$, where both spring and pendulum are stably at rest, is called the elliptical center. Oscillation around this center is referred to as elliptical or stable dynamics and is the only type of dynamics observed for a spring.

The situation is more complicated for a pendulum. In reality, a pendulum must be driven or "kicked"; otherwise, it slows down. The friction countering its motion has to be compensated for by a driving force. In reality, a pendulum dissipates energy. As the amplitude of the oscillation increases, the simple elliptical dynamics comes under the influence of a "hyperbolic center," where chaotic dynamics is observed.

Hyperbolic dynamics originates at the upper point of rest, $x = y = \pi$, where the pendulum is unstable. A slight jar is sufficient to start the pendulum rotating to the right or to the left of this position. Whether the pendulum passes through the point again or instead completes just one full oscillation is sensitively dependent on the initial conditions. This example clearly demonstrates the salient feature of chaotic dynamics: *the large consequences of small differences in the initial conditions.*

Elliptical orbits may be regarded as centers of *order* (stability, regularity). They describe the largely undisturbed orbits of planets and moons. Hyperbolic orbits may be considered centers of chaos (instability, unpredictability). They represent the orbits of comets, emerging suddenly from outside the solar system. If these centers are set side by side, their zones of influence interact in a very complex way. They interpenetrate and, even under a magnifying glass, there is no sharp boundary between them. At each stage of magnification, the competition between chaos and order is revealed anew.

The relative importance of chaos and order depends – in global terms – on the degree of nonlinearity of the respective system. As nonlinearity increases, the elliptical orbits increasingly vanish and chaos spreads out. The many narrow and at first hardly discernible bands of chaos merge into a few broad bands. Eventually, the large regions of chaos are separated by only a few boundary curves and, finally, the last of these vanishes as well. This last curve has something – almost mysteriously – to do with the *golden section*, often referred to as the *golden mean*. This astonishing fact, recognized only recently, has probably been responsible for the worldwide interest of mathematicians and physicists in the properties of such nonlinear maps. Indeed, the boundary separating order and chaos reveals an element of harmony. The key to understanding all this is provided by the so-called winding numbers. Depending on whether they are rational or irrational*, the maps of linear systems show either periodic or chaotic behavior. In nonlinear systems, the degree of rationality or irrationality is important. The "most rational" orbits respond most sensitively to

* Irrational numbers are real numbers that cannot be expressed as a quotient of two integers, for example, the number π (3.141...) or $\sqrt{2} = 1.4...$; rational numbers are numbers like $2 = 4/2$ or $1/3 = 0.333...$

nonlinear perturbations, whereas the "most irrational" persist the longest as elliptical curves. And the most irrational number of all is the *proportio divina*, the golden mean $d = (\sqrt{5} - 1)/2 = 0.618\ldots$

The mathematicians H. Poincaré and Garrett Birkhoff were the first to recognize that rational trajectories break up into chains of islands or "archipelagoes", as shown in Figure 5.7.

These resonances possess the property that elliptical and hyperbolic centers alternate. Part of the originally periodic trajectory survives, therefore, as a stable quasi-periodic trajectory, while the other part becomes chaotic. The relative proportion of the two depends on the degree of nonlinearity and the winding number of the resonance.

The opposite holds for irrational trajectories. Even when perturbed, they have a chance of survival. However, J. Moser and V. Arnold[8] have shown that not all irrational winding ratios are equally irrational. The golden mean d or its close relatives ($1 + d = 1/d = 1.618\ldots$, $1 - d = d^2 = 0.382\ldots$, etc) may be regarded as the "most irrational" and should therefore resist the onset of chaos the longest. This has been confirmed in computer simulations and is independent of the details of the map. The same holds for true Poincaré maps, discussed in connection with the double pendulum. Apparently, an element of universal harmony is intrinsic to this property of the golden mean winding number. We will learn more about this in chapter 6.

It is almost as if numerology, something Kepler discounted long ago in his early work "Mysterium Cosmographicum," has again crept into the description of physical systems, this time through the back door. Kepler wanted to interpret the interplanetary separations as a natural consequence of divine plan; the orbital spheres of the then-known six planets were inscribed in the five regular Platonic solids – a model in which irrational winding numbers played an important, if unrecognized, role. He distanced himself somewhat from these ideas when his analysis of Tycho Brahe's precise observations of Mars forced him to do so. In mathematics since Poincaré, however, such number games have once again become professionally acceptable. The analysis of scaled diagrams has still not explained all noteworthy rational and irrational ratios in our solar system, but it has readmitted very old questions to discussion.

With respect to the role of resonances in our solar system, mentioned at the beginning of this discussion, we have learned the following: On the one hand, rational orbits are unstable and break apart upon perturbation. This explains the gaps in Saturn's rings or in the asteroid belt. On the other hand, provided that the chaos is not too strong, the disrupted rational orbits still take on special configurations, which remain stable as elliptical islands. These remarkable situations apply to individual heavenly bodies insofar as their orbits are stabilizable by slight frictional forces. The cosmos thus appears to behave in ways that are not unreasonable from an earth-bound point of view. Rationality and resonance reign wherever large masses of matter such as the planets and their moons describe specific orbits, whereas irrationality and chaos dominate where many small objects (the asteroids and rings of matter) come together[7].

Chaotic phenomena have recently been used to explain several unusual features of other heavenly bodies as well. The unmanned Voyager mission to the outer planets revealed that Miranda, a moon of Uranus, has a geologically formed surface marked by eruptions and the flow of icy material. This is inconsistent with the low temperature of Miranda (< 100 K). The most plausible explanation is the following: An unequal distribution of density in the interior of the moon caused it to rotate chaotically. A satellite rotating in a nonsynchronous manner is acted upon by tidal forces, which warm it and, in the case of Miranda, made it possible for geological change to occur. Ultimately, the corresponding resonances set in. The motion became periodic and the moon cooled once again[9].

Fractal Dimensions

The topology of one-, two-, or three-dimensional objects cannot always be described using classical, integral dimensions. In order to describe reality, fractal dimensions are required[10]. This is illustrated in Figure 5.8.

In (a) and (b) from top to bottom are a series of curves, called Koch curves, displaying an increasing number of turns. Each curve is a smoothed out section of the one immediately below it and a more intricate version of the one immediately above it. The symmetry properties relating all these curves is termed self-similarity.

What happens when the curves are drawn with increasing fine detail while preserving self-similarity? If the dimension of this kind of object is described by defining it in terms of the number of "spheres" needed to cover the object[11], then it turns out that the lowermost curve in Figure 5.8 requires considerably more spheres to cover it than the uppermost. If two sections are compared, the lengths of which are in a ratio of $1:2$ in the uppermost curve, then the areas of the corresponding sections of the lowermost curve would behave as $1:2.88$, that is, $1:2^{1.5}$. Its dimension is therefore 1.5. The "curves" increasingly resemble a surface rather than a simple line.

The concepts of fractal dimension and self-similarity are essentially mathematical. For real physical and chemical objects, diffusion curves, the surfaces of crystals or of proteins, self-similarity is never satisfied ideally over the entire linear scale. There exists an upper and a lower limit. When a surface is broken down ever further into self-similar fragments, they become increasingly fissured and higher-dimensional. This process reaches its maximum, and also its limit, when molecular dimensions are attained. But there is also a limit in the macroscopic direction. Viewed macroscopically, a mirror has a dimension of exactly two. If the surface is viewed under

Fig. 5.8. Koch curves with fractal dimension $D = 1.5$ (a) and $D = 1.79$ (b)[11].

the electron microscope, on the other hand, it most likely resembles a mountainous landscape of higher dimensionality. The table shows examples of fractal objects in nature. As can be seen, the surface of a protein, for example, does not have the classical dimension 2.0, but rather 2.4. This is undoubtedly of importance for the precise mathematical treatment of protein interactions, for the mechanisms of enzyme action, and for all life processes involving biopolymers.

A special role is played by fractal dimensions when the surface of a catalyst is covered by molecules, as illustrated by the so-called Menger sponge (Fig. 5.9). The catalytic properties of this surface vary widely, depending on its dimensionality. Fractal

Object	Fractal dimension
Coastline	1.2
Landscape	2.2
Surfaces of clouds – formed experimentally as well as theoretically by chaotic dynamics (turbulence)	2.35
Cross-linked polymers, gels	2.5
Chain polymers in good solvents	1.67
Brownian motion in two and three dimensions; molecular trajectories in liquids	2
Energy levels of molecules	<1
Skeletons of proteins	1.3 – 1.8
Surfaces of proteins	2.4
Surfaces of solids	2 – 3

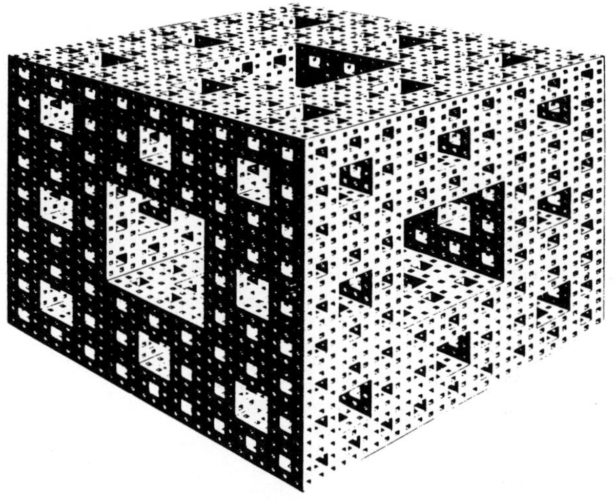

Fig. 5.9. Menger sponge ($D = \ln 20/\ln 3 = 2.73$) as a model for a porous catalyst (all centers of the "cubes" are empty). If it is covered with a molecule of size r_0, the resulting monolayer possesses a well-defined surface (for resolution r_0), total length of the edges, and number of corners. Division of these three magnitudes by r_0^2, r_0^1, and r_0^0, respectively, defines the number of (effective) surfaces, edges, and corners on the surface. The sponge further illustrates that a molecule diffusing along the surface can move from one point to another through the pores via a much shorter pathway that it can on a nonporous surface of the same dimension. The resulting fractal dimension is $D>2$ (see p. 126).

dimensions (cf. Fig. 5.10) also play an important role in the reaction kinetics of quasi-periodic or chaotic chemical processes, such as the Belousov-Zhabotinskii reaction (cf. chapter 1). Moreover, the trajectories of coupled systems, such as those discussed

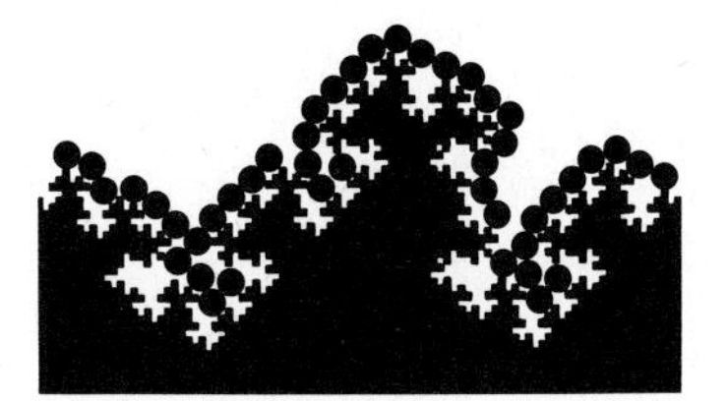

Fig. 5.10. The surface of a catalyst covered with a monolayer of different-sized molecules (schematic).

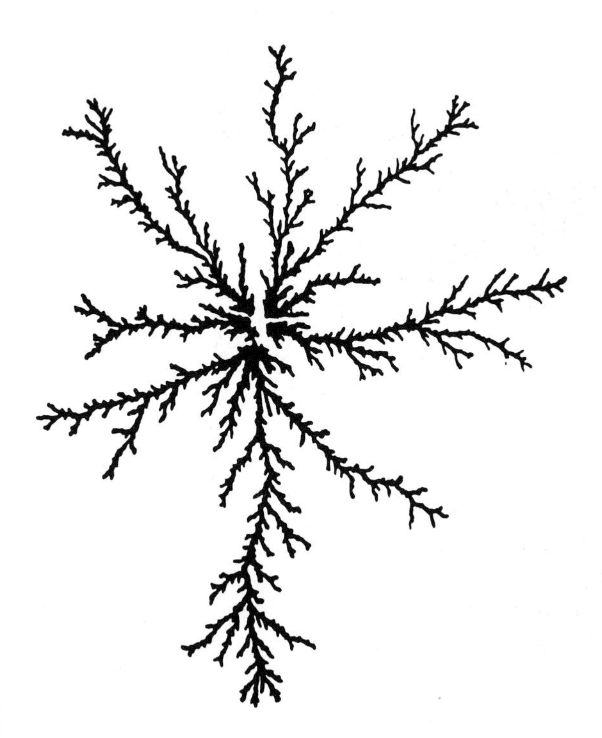

Fig. 5.11. Electrolytically generated (planar) zinc tree. Optical measurement (essentially mass-radius relation) gave $D = 1.66 \pm 0.03$. This value is in very good agreement with the dimension of the dendritic structure formed by diffusion-controlled growth (computer simulation)[12].

for the three-body problem, are measured with fractal dimensions. Indeed, wherever chaos is present, the dimensions become fractal. We have thus come full circle from our discussion on lightning, the Lichtenberg figures, and tree structures. These, too, are fractal. The electrolytically generated zinc tree shown in Figure 5.11 is in principle a one-dimensional drawing, but in fact has a fractal dimension of 1.6.

Wallace Stevens

Connoisseur of Chaos

I

A. A violent order is disorder; and
B. A great disorder is an order. These
Two things are one. (Pages of illustrations.)

II

If all the green of spring was blue, and it is;
If all the flowers of South Africa were bright
On the tables of Connecticut, and they are;
If Englishmen lived without tea in Ceylon, and they do;
And if all went on in an orderly way,
And it does; a law of inherent opposites,
Of essential unity, is as pleasant as port,
As pleasant as the brush-strokes of a bough,
An upper, particular bough in, say, Marchand.

III

After all the pretty contrast of life and death
Proves that these opposite things partake of one,
At least that was the theory, when bishops' books
Resolved the world. We cannot go back to that.
The squirming facts exceed the squamous mind,
If one may say so. And yet relation appears,
A small relation expanding like the shade
Of a cloud on sand, a shape on the side of a hill.

IV

A. Well, an old order is a violent one.
This proves nothing. Just one more truth, one more
Element in the immense disorder of truths.
B. It is April as I write. The wind
Is blowing after days of constant rain.
All this, of course, will come to summer soon.
But suppose the disorder of truths should ever come
To an order, most Plantagenet, most fixed...
A great disorder is an order. Now, A
And B are not like statuary, posed
For a vista in the Louvre. They are things chalked
On the sidewalk so that the pensive man may see.

V

The pensive man ... He sees that eagle float
For which the intricate Alps are a single nest.

6. The World Is Harmonic

Heinrich von Kleist* recounts a conversation with the maître de ballet C. on puppet theater [1]

The winter of 1801 I spent in M...one evening I encountered Herr C. in the public garden. He had recently been engaged by the city as the principal male dancer in the opera and was enjoying a raging success with the public.

I told him I was surprised to have seen him several times at a puppet theater constructed in the marketplace, which entertained the public with short dramatic burlesques interlaced with song and dance.

He assured me that the pantomime of these marionettes gave him great pleasure, and he stated in no uncertain terms that a dancer seeking to train himself could learn much from them...

He asked me whether I didn't also find some of the movements of the dancing marionettes, especially the smaller ones, quite graceful. I had to admit he was right...

I inquired about the mechanism of these figures and asked him how on earth their separate limbs and joints responded to produce the rhythms required by the movements, or the dance, without myriads of strings being attached to one's fingers?

He said I shouldn't be misled into imagining that, at every moment in the dance, each limb was positioned and jiggled separately by the puppeteer.

Each movement, he said, had a center of gravity and it sufficed to manipulate this from within the figure; the limbs, which were really nothing more than pendulums, responded automatically, mechanically. Nothing further was required.

He added that this motion was very simple and that, each time the center of gravity was moved in a straight line, the limbs described curves. Even a slight accidental jiggle was enough to set the whole into a kind of rhythmic movement, resembling a dance.

This remark seemed at first to shed some light on why he professed to be entertained by the puppet theater. At the time I had no idea of the conclusions he would later draw from all this.

---

* German poet and writer, 1777–1811.

I asked him whether, in his opinion, the puppeteer was himself a dancer or at least had some feel for the beauty of dance.

Merely because a job was mechanically simple, he replied, did not necessarily mean that it could be performed without any feeling whatsoever.

The line traced by the center of gravity was simple and, he thought, usually straight. Even when it was curved, the equation describing its curvature was at least first order and at most second. And in the latter case, it was merely elliptical, a form of motion entirely natural for the extremities of the human body (owing to the joints). The puppeteer didn't have to be a great artist to accomplish this. Yet, from another point of view, this line was quite mysterious. For it was nothing less than the path followed by the dancer's soul. And he doubted that it could have been found at all unless the puppeteer had imagined himself at the center of gravity of the marionette, that is, unless he himself had danced.

I responded that I had imagined this business to be rather mindless, much like turning a crank to play a barrel organ.

On the contrary, he retorted, the movements of the puppeteer's fingers were related to those of the attached marionette in a rather sophisticated way, like the relationship of numbers to their logarithms or of the asymptote to the hyperbola...

I expressed my amazement at how he was able to appreciate this form of public entertainment as a high art. Not merely that he held it capable of higher development; he seemed to be involved in it himself...

He smiled and said that he would go so far as to assert that, if a craftsman were willing to construct a mechanical figure to the specifications he had in mind, he could use it to perform a dance that neither he nor any other accomplished dancer of his time would be capable of imitating.

No matter how clever he was at realizing this paradoxical situation, I said, no one could make me believe that a mechanical marionette could embody more grace than the human physique.

He retorted that, in this respect, man was incapable of coming even close to a marionette. Only God could compete with matter in this field of endeavor. And this was the point at which the two ends of the ringlike world interlocked.

I was growing more and more perplexed and didn't know how I should respond to such extravagant statements.

It seemed, he added, as he took a pinch of tobacco, that I had not read carefully enough the third chapter of the First Book of Moses, and that whoever lacks an understanding of this first period of all human learning is incapable of discussing the following periods, let alone the final one.*

I said that I was perfectly well aware of the disorder

human consciousness had inflicted on man's natural grace. One time, before my very eyes, a young man with whom I was acquainted lost his innocence through a

---

* Eating from the tree of knowledge.

casual remark and never regained this paradise, his every effort notwithstanding. – Well, I added, what conclusions could you draw from this?

He asked me what incident I was referring to.

About three years ago, I was bathing with a young man whose natural physique, at the time, radiated a wondrous aura of grace. He was about sixteen years of age. Only faintly did he reveal the first traces of vanity, induced by female attention. It so happened that, a short time earlier, we had been in Paris and had seen the famous statue of the youth removing a splinter from his foot, casts of which are found in most German collections. Glancing into the large mirror as he placed his foot on the stool to dry, he suddenly recalled the statue; he smiled and said to me that he had made quite a discovery. Indeed, at that very moment, I had made the same one; but, for some reason – whether to test the genuineness of his inherent grace or to counter his vanity in a healthy manner – I smiled and replied that he was just imagining things. He blushed and then raised his foot a second time in order to show me; but the attempt, as might have been anticipated, failed. Flustered, he raised his foot a third and fourth time; he surely must have raised it ten more times; in vain! He was incapable of producing the same movement. What am I saying? The movements he made had such a comical element that I had to stifle my laughter.*

From that day on, indeed from that very moment on, the young man was overcome by an unforeseeable change. He began to stand for days on end in front of the mirror, gradually losing his charm. An invisible and unimaginable power seemed to cast its iron net over him and to restrain the free play of his gestures. After one year had passed, there was no longer any trace of his original charm, which the people around him had once idolized. There is someone still alive today who witnessed this strange and unfortunate string of events and can attest to what I have said, word for word.

I would like to take the opportunity, said Herr C. in a friendly tone, to tell you another story. You will quickly realize why it is relevant here.

During my trip to Russia, I stayed at the country estate of Herr von G.,. . .a Livonian noble, whose sons were intensively engaged, just then, in fencing practice. Especially the older one, who had just returned home from the university, sought to display his skill and, one morning when I happened to be in his room, he offered me a rapier. We fenced; but it so happened that I was superior to him; furthermore, I became keen to fluster him; nearly every one of my thrusts struck him and finally his rapier flew into the corner. After recovering the rapier, he said – half joking, half serious – he had met his match, but everything in the world finds its match and, therefore, he wanted to lead me to mine. The brothers laughed loudly and cried: Let's go! Let's go! Down to the stable! And they took me by the hand and led me to a bear, which Herr von G. . ., their father, was raising in the courtyard.

---

* He is referring to a statue in the Louvre, the Hellenistic sculpture of a young man removing a thorn from his foot.

Astonished, I approached the bear, which raised itself on its hind legs, leaning its back against a post to which it was chained. It raised its right paw, ready for combat, and stared me in the eye; that was its fencing stance. Standing face to face with such an opponent, I asked myself whether I was dreaming. Thrust! thrust! said Herr von G. . ., and try to teach him a thing or two! Having recovered somewhat from my first astonishment, I plunged at him with the rapier. But the bear just moved his paw a bit to the side and parried the thrust. I tried to confuse him by feinting; the bear didn't move. I thrust again at him with such momentary dexterity that I would have struck a man's chest for sure; the bear made a slight movement with its paw and parried the thrust. Now I had almost fallen in the trap set by the young Herr von G. . .. The bear's gravity was coming close to robbing me of my composure. I began to drip with sweat as thrust and feint alternated. In vain! Not only did the bear, like the best fencer in the world, parry all my thrusts; it did not even respond to feints, something no fencer in the world could imitate; eye to eye, as if it could read my very soul, it stood there, its paws raised. When my thrusts were only feigned, it didn't budge.

Do you believe this story?

Absolutely! I shouted with enthusiastic applause – from the mouth of any stranger, so real is the tale, and from yours all the more.

Now, my dear friend, said Herr C. . ., you have all you need to understand me. We see that, to the extent that the reflections become darker and weaker in the organic world, the grace therein emerges with ever more radiance and majesty. Just as the intersection of two lines on one side of a point recedes to infinity on going through that point and then suddenly reappears right in front of us, so, too, does grace reemerge after knowledge has gone, as it were, through infinity; it thus assumes its purest form in the physique with either no consciousness or an infinite consciousness, that is, in the marionette or in God.

Does this mean, I said, somewhat absentmindedly, that we have to eat once again from the tree of knowledge in order to revert to the state of innocence?

Absolutely, he answered; that is the final chapter in the history of the world.

The Harmony of the Spheres – Kepler Was Right After All

In his work "Mysterium Cosmographicum," which appeared in 1594, Johannes Kepler attempted to explain the planetary orbits in terms of supreme harmonies. According to his system, the planets move on the surfaces of spheres in which the Platonic solids – tetrahedron, cube, and pentagonal dodecahedron – are inscribed (Fig. 6.1).

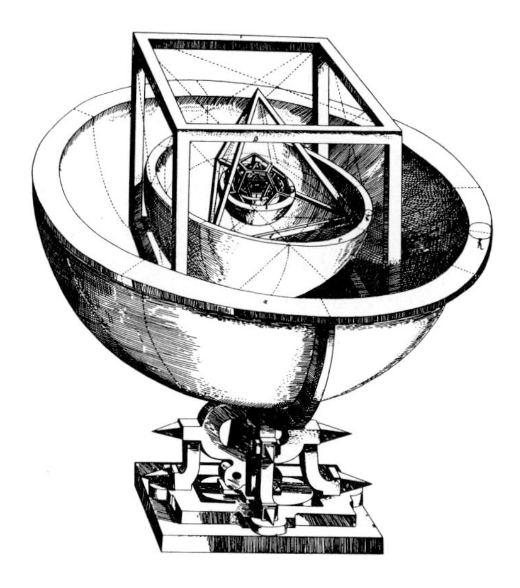

Fig. 6.1. Kepler's model of the solar system. The planets move on the surface of spheres in which the Platonic solids are inscribed. In this way, Kepler tried to explain the interplanetary distances.

When the young Kepler tried to unify exact science and divine mystery in "Mysterium Cosmographicum," his thoughts and feelings were entirely consistent with the mystical tradition of the Middle Ages: "Great is our Lord and great is His might, and His wisdom knows no end. Sing praises unto Him, sun, moon, and planets, in whatever language your song of praise be heard by the Creator. Praise Him, you heavenly harmonies, and also you, those who witness and confirm His revealed truths! And you, my soul, sing the glory of the Lord throughout your life! From Him and through Him and to Him are all things, the visible and the invisible. To Him alone are honor and glory from eternity to eternity! I give thanks to You, Creator and Lord, that You have granted me this joy in Your creation, the rapture in the works of Your hands. I have proclaimed the majesty of Your works to all the peoples, insofar as my finite spirit is able to comprehend Your infinity. Where I have said something unworthy of You, or where I have striven after my own honor, then forgive me in mercy."[2]

The ancient yearning to relate the structure of the world to underlying harmonies is documented at least since Pythagoras. He believed in the "harmony of the spheres," an intrinsic harmony of the universe that man, lacking the proper sense organs, cannot perceive directly. A divine or enlightened being, on the other hand, is capable of hearing wondrous harmonic sounds due to the coordinated motions of the heavenly bodies. Kepler later abandoned or modified these mystical notions when he set forth his laws based on the more exact observations carried out in the meantime by Tycho Brahe.

Kepler's laws are well-known:

1. The planets move in ellipses with the sun at one focus.
2. The lines joining the center of the sun with each planet sweep out identical areas in equal times.

3. The squared periods of the planets are proportional to the cubes of their mean distances from the sun.

The very discovery of these laws represented an astounding act of intuitive "harmonization." Complex, nearly unmanageable data could be sorted out by applying these empirical generalizations without the necessity of invoking a higher law. And Kepler's laws are really not much more than empirical generalizations. They became laws only within the context of Newton's more encompassing laws of motion and they were incorporated into Newton's theory of gravity.

In this sense, Kepler's laws – indeed, every new scientific theory – may be regarded as a milestone on the path to understanding the harmony of the world. Isn't this how we should regard all physical laws? Even those that seem to make the world more complicated. Does the structure of our mind predispose us to seek only harmonies in the world and to ignore disharmonies, which, in the world as a whole, counterbalance the harmonies? Or is the world intrinsically harmonic, so we have no choice but to discover harmonies? After all, since our mind is also a part of the world, the forms of our knowledge should reflect the structure of the world. Or expressed in a different way: Is our knowledge acquired by filtering out the (few) harmonies, or is cognition an act corresponding to the structure of the world? This question of transcendental philosophy cannot be answered objectively. But if we take *ourselves* seriously and if we take *the world* seriously, we are compelled to acknowledge that the deeper we probe into the interrelations the more harmonies we discover.

But back to Kepler. We have already seen in chapter 5 that almost mystical numerical ratios are established in the planetary system. The interaction of heavenly bodies gives rise to resonances. The periods of Jupiter and Saturn are in a ratio of almost exactly 2:5. In the asteroid belt, the numerous tiny planets between Jupiter and Mars, there are gaps where asteroids would have had periods of one-half, one-third, or one-fourth the period of Jupiter (cf. chapter 5, reference 5). Only recently have we begun to understand these numerical ratios – at first mystical and somehow reminiscent of Pythagorean numerology – by analyzing them in terms of many-body systems coupled by feedback. Kepler's first, but later-revoked, work was therefore correct after all: Harmonies are established all by themselves in this world when, under certain feedback conditions, chaos is allowed to "play itself out."

Mandelbrot Figures – The Beauty of Fractals

In recent years modern mathematics has focused increasingly on complex feedback processes. Such processes have long been known in principle and are describable in

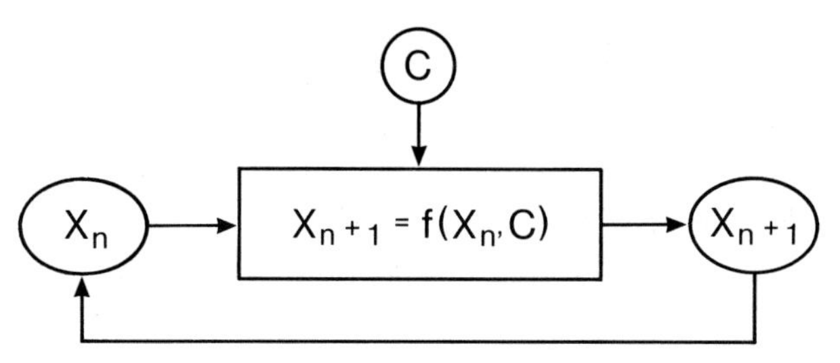

Fig. 6.2. Complex feedback system. This process is formally the same as that involved in the baker transformation (Fig. 4.19), in enzymatic feedback (Fig. 1.11, 1.12, or 4.12), the dynamics of growth (Fig. 6.5), or the error catastrophe (Fig. 8.4).

simple cases by differential equations, according to the principles developed by Newton and Leibniz. The laws of dynamics thus allow one to determine the trajectories of moving bodies or systems if these trajectories are regarded as continuous paths or treated stepwise. This is the essence of differential and infinitesimal calculus.

Feedback governs virtually every living process and can be ignored only in crude simplifications[3]. This feedback character of nearly all systems in nature is illustrated in Figure 6.2.

In reality, though, a *nonlinear* relation exists between the input X_n and the output X_{n+1}. In other words, the law $X_{n+1} = f(X_n, C)$ is more complex than the simple proportionality $X_{n+1} = KX_n$. The relation is dependent on the value of C, which reenters the process during each iteration. For an arbitrary initial value X_0, one wants to know what happens to X in this process (the overall operation has to be regarded as a process).

There are three basic possibilities. First, the final value of X approaches a limiting value, which it ultimately attains, perhaps asymptotically (at infinity). This is the case of linear differential equations and integrable systems. Second, the process leads to harmonic oscillation. This case corresponds to the dynamics of the pendulum and the planetary orbits. Third, the outcome of the process is uncertain. Although the process is governed by its dynamics and the choice of initial conditions, its outcome is nonetheless unpredictable. All three situations are possible in physical and physiological reality. We already saw in chapter 5 that chaotic bands are present even in our "simple" planetary system. The three-body problem, represented by the double pendulum, also yields chaotic solutions, This is shown once again on the left-hand side of Figure 6.3.

In reality, most systems are hybrid systems with some chaotic solutions. A famous example is the Lorenz attractor (Fig. 6.3, right), which describes dissipative structures formed when a mechanical system is dampened by a second center of attraction through friction or other forms of energy loss. Examples include the double pendulum and the tides. Ultimately, the trajectory "plunges," so to speak, into one of the "gravitational centers." However, this trajectory is not a simple spiral, but springs

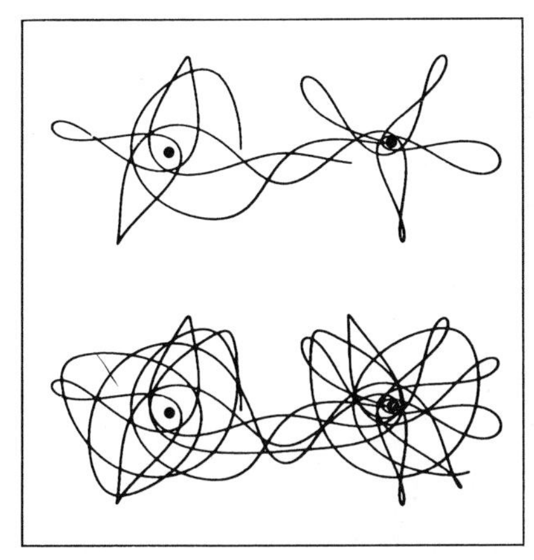
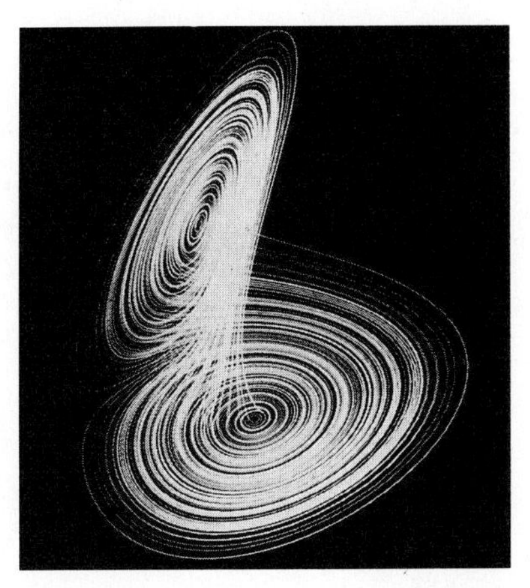

Fig. 6.3. Left: Schematic representation of a small planet in orbit around two suns of identical mass. Top left: Initial trajectories. Bottom left: Later trajectories of the chaotic motion. Right: the Lorenz attractor.[3]

back and forth between the two centers of attraction. These systems are referred to as strange attractors. The visual representation of such trajectories has great aesthetic appeal – in short, it is beautiful. Could it be that what we find beautiful is, in fact, the juxtaposition of order and chaos?

The basic equation of Figure 6.2 is represented in a different way in Figure 6.4, which shows a Mandelbrot set surrounded by the Julia sets it controls. These figures are generated by solving the feedback equation $X_{n+1} = X_n^2 + C$, where C is a complex constant. The Mandelbrot figure has fractal dimensions at its chaotic boundaries. As the boundaries are viewed at higher and higher resolution, the Julia sets are revealed in new and ever-changing detail. We cannot go into the mathematical treatment of such systems in this discussion. The interested reader is referred to the original publications[4,5,6].

How do fractal structures and chaotic boundaries arise? Are there real examples of such states? Are these just mathematical games or are they really required to describe actual states? Let us take an example from population dynamics: the growth of a population of flies, rabbits, or human beings. Under favorable conditions, a rabbit population, for example, increases until the biotope is filled with rabbits and the number of rabbits exponentially approaches its saturation limit (Fig. 6.5).

In a paradisiacal world, the population would remain constant at this saturation value. But reality is no paradise. Predators such as foxes prey on the rabbits. The population of foxes then starts to increase as well, although this increase is obviously delayed by several generations. The foxes gorge themselves and decimate the rabbit population to near extinction. Not being very environmentally minded, they fail to notice that their food resource is vanishing. The ensuing famine decimates the fox

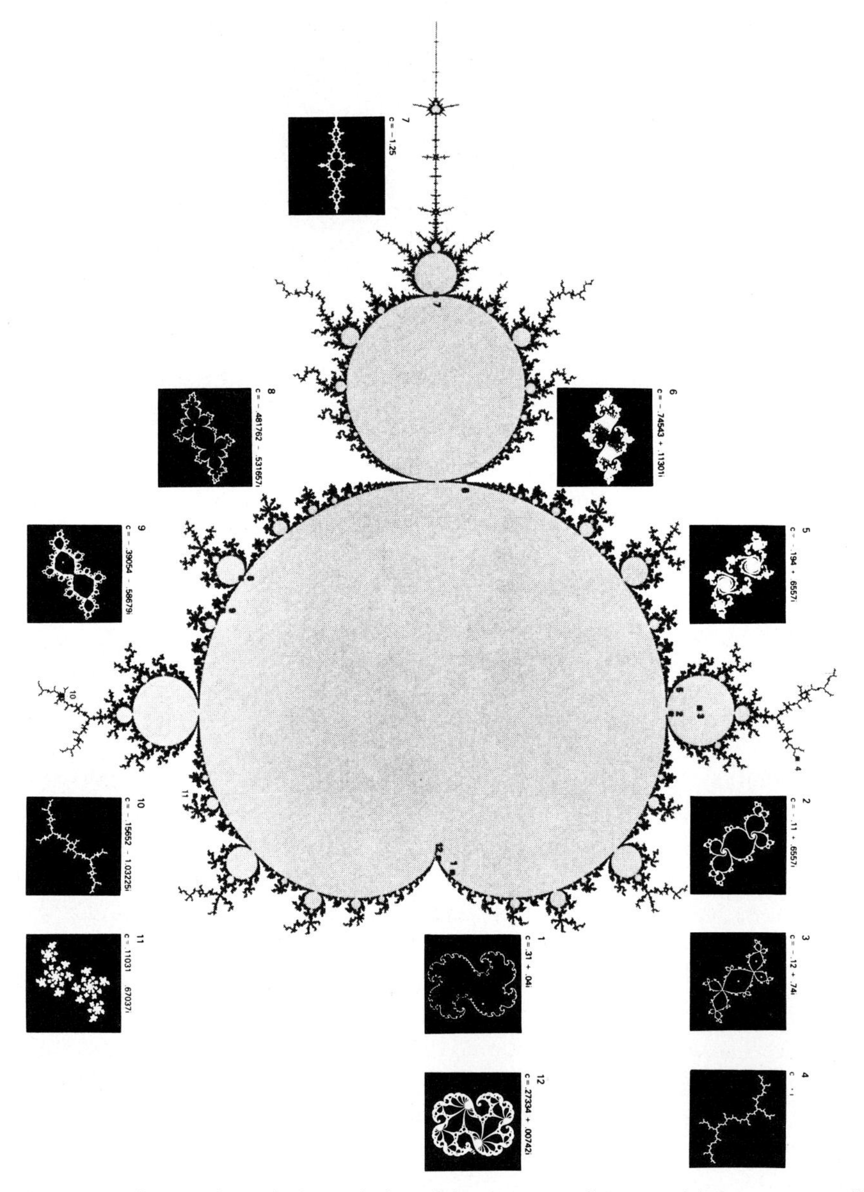

Fig. 6.4. Julia sets along the boundaries of the corresponding Mandelbrot set controlling their structure.[3]

population, and, few years later, the population of rabbits is again on the rise. The overall process thus exhibits periodic feedback. In 1845, B. F. Verhulst derived a growth law for processes in which growth occurs at a variable rate subject to feedback and is therefore nonlinear. This nonlinearity has unexpected consequences for

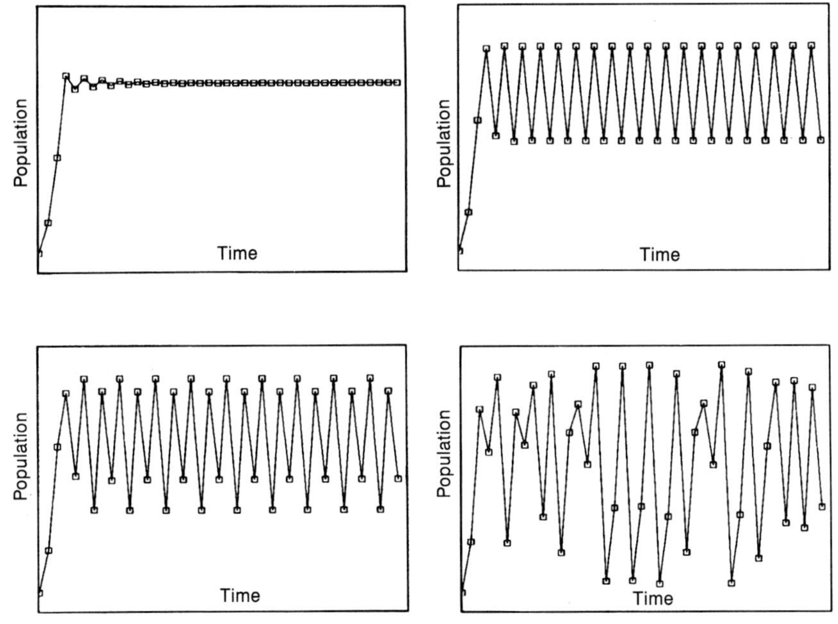

Fig. 6.5. Verhulst growth dynamics.

the dynamical behavior, especially when the growth rate is very high – say, greater than 200 percent. The conceivable situations are illustrated in Figure 6.5. The plot at the top left shows how the limiting value is normally attained for a growth rate of 180 percent ($r = 1.8$; $r = 1.0$ would correspond to a steady-state population, $r < 1.0$ to negative growth). At 230 percent, the system starts to oscillate (top right), the rapid growth causing the system to exceed the limiting value x. For a growth of 250 percent, the oscillation splits in two (period doubling, bottom left) and, finally, chaos results for a growth of more than 257 percent (bottom right). Although these growth rates are not very realistic for higher forms of life, they are quite common for insects and microorganisms. In symbiotic processes, moreover, they are of potential importance for man as well. For instance, all diseases occurring as periodic attacks or outbreaks are probably best understood in this way. Examples include tertian or vivax malaria, which has a three-day rhythm, periodic infestations of insect pests, and the specific incubation times of infectious diseases. During the past twenty years, the Verhulst law has found particularly important application in meteorology and in many branches of physics. Some of these applications were discussed in the preceding chapter.

The Verhulst law can be formulated as follows:

$x_{n+1} = (1+r)x_n$, where n is a unit of time (the number of years, the generation number, etc.) and r is the growth rate. Since the size of the population affects the growth rate through feedback coupling, however, the equation assumes the form $x_{n+1} = (1+r)x_n - rx_n^2$.

What does it mean to say there is chaos for a growth rate of over 257 percent? This means that, for strictly deterministic initial conditions, indeterministic "jumps" occur. This is shown in the bottom right of Figure 6.5. In many respects the figure resembles the baker transformation discussed earlier, which essentially describes a growth process, namely, stretching something to twice its length at each "generation." Verhulst dynamics makes it possible to specify more precisely the transition from order to chaos. Where exactly are the bifurcation points at which oscillation, frequency doubling, and chaos appear?

Period doubling (Fig. 5.6b and c) and the transition to chaos (Fig. 6.5d) set in at particular points. Let us define r_n as the value of the growth rate corresponding to the nth bifurcation (doubling, quadrupling, etc., until chaos sets in). Then the length of two successive bifurcation events is

$$\delta_n = \frac{r_n - r_{n-1}}{r_{n+1} - r_n}$$

In our very crude example, this would be

$$\delta_n = \frac{2.50 - 2.30}{2.57 - 2.50} = \frac{0.20}{0.07} \approx 3.0$$

As the process proceeds further, deeper and deeper into chaos, as it were, this quotient approaches the value

$$\delta_n \to \delta = 4.669201660910\ldots$$

This number, called the Feigenbaum number after the American mathematician M. Feigenbaum, is a universal constant that describes the transition from order to chaos, just as the number $D = 3.1415926536\ldots$ describes the ratio of the circumference of a circle to its diameter. This constant was discovered by the German mathematician S. Grossmann (born in 1930). Feigenbaum later characterized it in greater detail.

It, too, is an irrational number; that is, written in decimal form, it neither comes to an end (no matter how many decimal places are given) nor shows periodicity. What an astounding discovery! By analyzing in detail an equation applicable to biological systems – the Verhulst equation – mathematicians and physicists found a universal constant that describes abrupt transitions throughout nature, be they periodic changes, period doubling, or transitions from periodicity to chaos (see also chapter 1, Fig. 1.11). The compelling conclusion is that chaos is an intrinsic feature of natural systems. In other words, the basic structure of the world is nonlinear, even though islands of order, where our simple linear laws are still applicable, continually emerge out of this deterministic chaos. The linearization that we have to perform in the Cartesian-Newtonian system in order to arrive at physical laws at all is thus insular. This is revealed in particular at the "shores" or boundaries of these islands (cf. Fig. 6.4).

During the last decade or so, the mathematician Benoit Mandelbrot[4] has investigated a number of feedback equations in which imaginary numbers appear and in which the parameter C of our feedback equation is an imaginary or complex number[5,6]. (Complex numbers are numbers that contain real – for example, 2 – and imaginary factors – for example, $\sqrt{-1}$. An example is $2+\sqrt{-1}$. $\sqrt{-1}$ is an imaginary number because it should not really exist at all. Since -1 times -1 or $[-1]^2$ is $+1$ [$-$times$-$gives $+$], -1 cannot have a "real" square root.) Such equations or, better, sets (in the mathematical sense) are best visualized on the display screen of a computer. By applying a few general rules to the Verhulst equation or the Julia set, it is possible to obtain displays of exquisite complexity and stunning beauty. These are pictorial representations of feedback equations – in this case with imaginary or complex numbers – and thus represent the results of simple calculations based on a feedback procedure without any harmonizing or aesthetic constraints. The underlying mathematics is a form of description adapted to a nonlinear reality. It fits the reality much better than the artificial abstractions of Newtonian trajectories. And this mathematics leads, in visual terms, to pictures of supreme harmony. The world is harmonic.

Why Is Nature Beautiful? On Blossoms and Fruits

Many phenomena in living nature we find beautiful, harmonic, symmetric. Why do certain calyxes have five sepals? Why are animals basically symmetric in form? Since these structures are the products of growth, could they not have arisen just as well in a chaotic and uncontrolled manner, like cancer cells?

Since antiquity, starting with Pythagoras, philosophers and natural scientists (like Johannes Kepler) have devoted their interest to naturally occurring symmetries and proportions. The intriguing fact emerged that natural proportions often obey what is referred to as the golden section or golden mean. In Figures 6.6 and 6.7, the forms of several kinds of leaf and snail are shown with the "golden proportions" indicated. These sketches were taken from a book more than one hundred years old[7].

P. Richter and R. Schranner[8] have studied different plants, blossoms, and fruits from this perspective (Fig. 6.8a, b, and c show some examples). Figure 6.8a shows a fir twig whose surface pattern is highlighted with superimposed *helical* segments. Goethe[9] himself pondered the spiral tendencies in nature and described the phenomenon as follows: "The spiral system evolves, propagates, maintains itself, thus

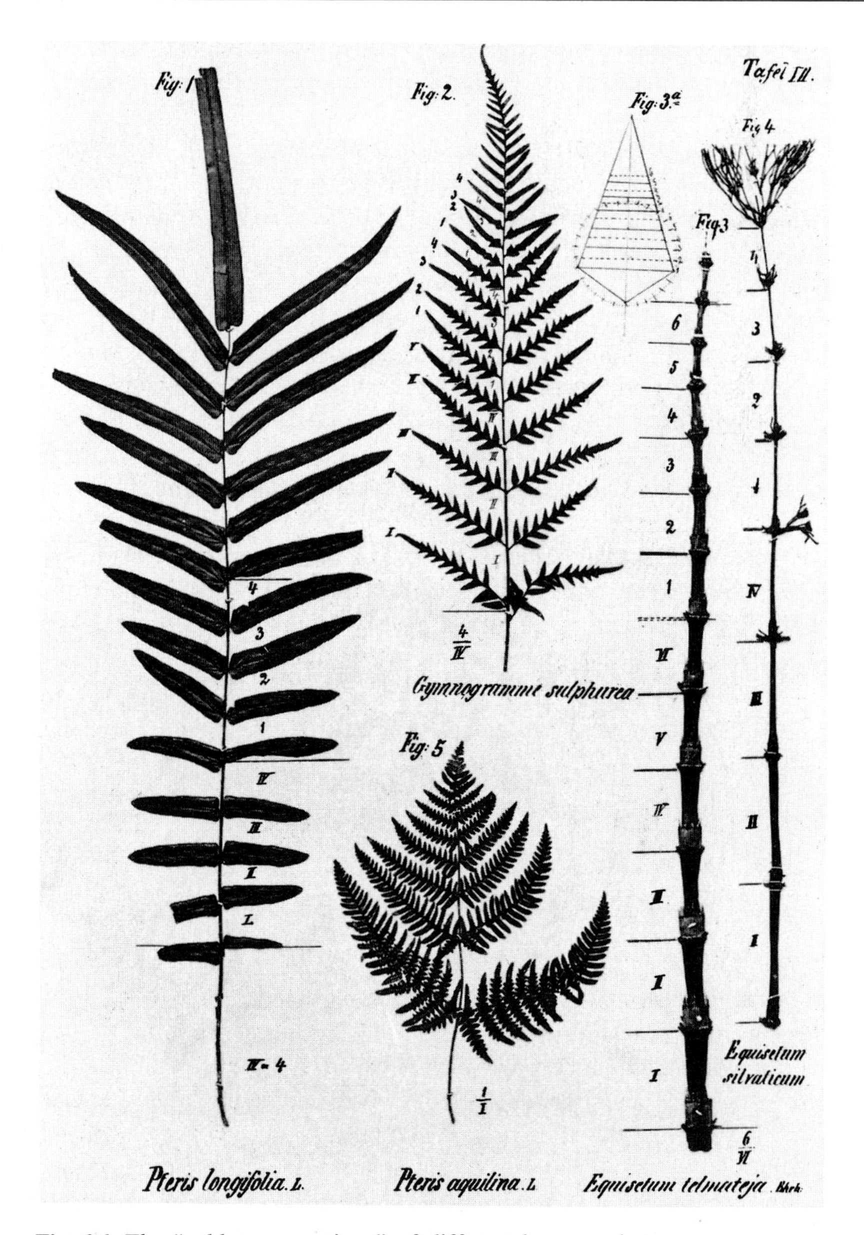

Fig. 6.6. The "golden proportions" of different leaves and stems.

bypassing and at the same time isolating itself from the vertical system. Overextending itself, it very soon becomes frail and is subject to decay; if it is combined with the vertical system, both grow to form a lasting entity as wood or some other solid. Neither of the two systems can be imagined alone. They are always and eternally to-

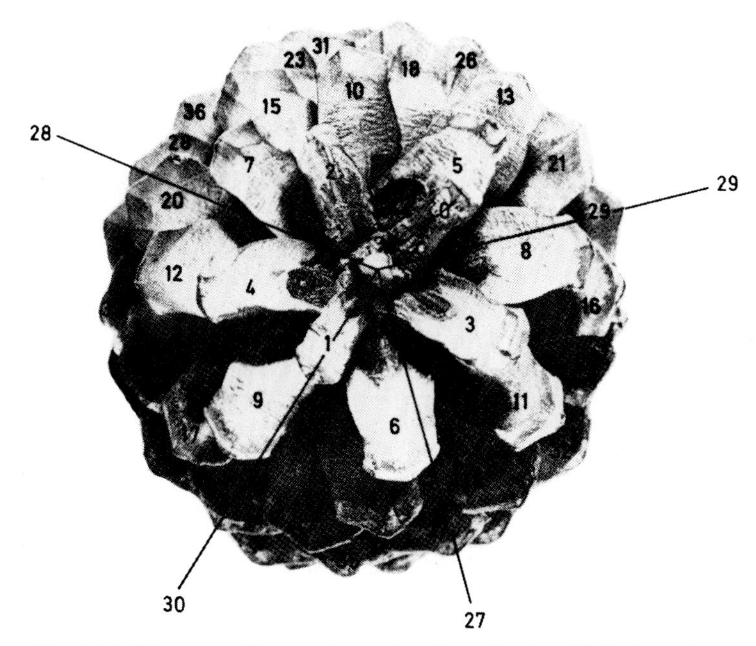

Fig. 6.10. Pinecone. The numbers give the order in which the scales are formed.

of a circle according to the ratio of the golden section: $\Phi/(2\pi - \Phi) = (2\pi - \Phi)/2\pi$. The golden angle thus ensures that the spiral always possesses Fibonacci character, independent of the particular conditions of growth.

Whereas the growth occurs outward from a center for spirals in a plane (composite blossoms of the sunflower), it occurs along an axis for a *helical* stem. In both cases, the golden angle or the Fibonacci series is maintained with respect to the number of intersections in the individual spirals. In fact, this spiral tendency reflects a general law of nature. Figure 6.11 shows the generation of Fibonacci patterns in a composite flower.

Each kind of growth can be regarded as a phenomenon of internal competition. Certain activators or growth factors induce a particular element or bud to grow. At the same time, this growth is inhibited by competing processes in the immediate vicinity. This inhibition may be caused by natural inhibitors produced through metabolic processes at the point of growth or by competition for the nutrients required for growth[10].

Here, too, feedback systems are at work; two effects either compete with or support each other. After the formation of the first bud or leaf or, in the case of composite flowers, after the formation of the pistil, the next budding takes place opposite and in the most distant position possible. Naturally, this cannot occur outside the "system of the blossom." Opposite means at an angle of 180 degrees (Fig. 6.12, posi-

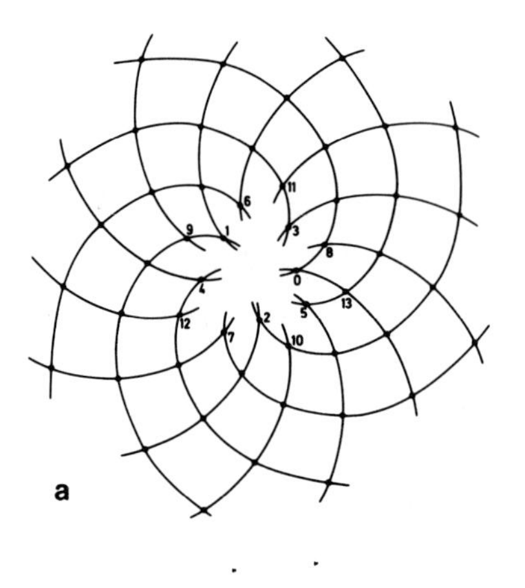

Fig. 6.11. Generation of Fibonacci patterns in a composite flower; the coordinates of the intersections obey the formula $(X_k, Y_k) = (r_0 + k\Delta r)$ cos $k\Phi$, sin $k\Phi$). X and Y are the coordinates of the point being formed, r is the rate of growth, and Φ is the "golden angle." Rapid growth (a) results in relatively low Fibonacci numbers (8/13), more compact packing (b) in Fibonacci numbers of 13/21; (c) slow growth. The regular Fibonacci spirals of 5 to 144 are indicated; the broken spirals are deviations from the Fibonacci rule due to some disturbance of normal growth.

b

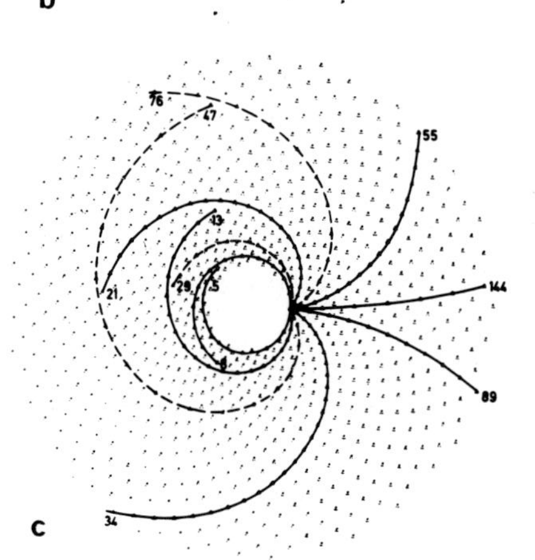

tion 2). Where does the next bud form? The inhibitory effect originating at position 2 in Figure 6.12 will still be large, whereas that at position 1 will have decreased substantially. The bud therefore forms closer to position 1 (position 3). The same holds for the formation of the fourth bud and so on. If one assumes an exponential de-

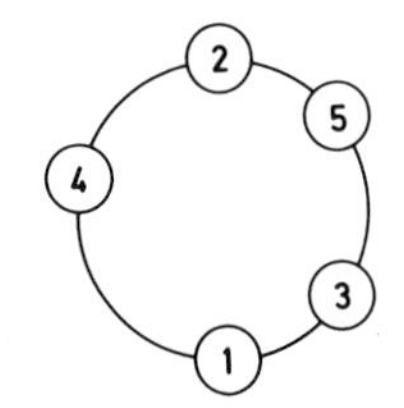

Fig. 6.12. Bud formation governed by inhibitors.

crease in the growth tendency of the individual buds with time, in accordance with observation, then the following formula holds: $i_3 : i_2 = i_2 : i_1$, where i is the "inhibition strength" of each point at which budding occurs. Accordingly, the effect of the penultimate bud on the growth of the previous bud cannot be completely ignored. Straightforward calculation yields the golden angle.

The simple assumption made above clearly involves a number of unknowns. It is still not known with certainty what "growth pressure" and "inhibition by neighbors"

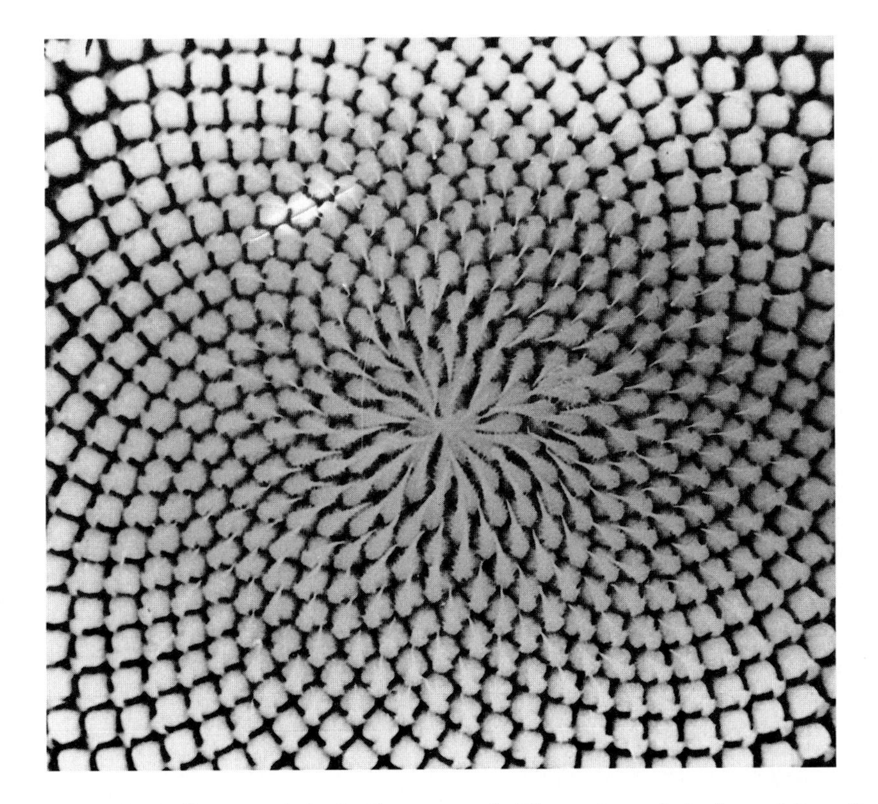

Fig. 6.13. Sunflower seeds in the flower capsule. The seeds are fruits formed from the pollinated pistils. The sites where the flower pistils bud are governed by the feedback principle illustrated in Figures 6.11 and 6.12. Fibonacci spirals are clearly visible and easily counted. Isn't this pattern beautiful?

really mean. Do they imply biochemical growth and inhibition factors? Is the inhibition due to the physical pressure of neighboring buds? Whatever the nature of the biochemical process, it undoubtedly involves feedback; thus, the growth of beautiful blossoms obeys the rules of the golden section. The fascinating pattern of sunflower seeds is shown once again in Figure 6.13.

The golden section is generated whenever the longer part is to the whole as the shorter part is to the longer: $d:1 = (1-d):d$. This situation yields an irrational number, a number that cannot be expressed as the quotient of two integers – for example, the square root of any prime number. In decimal form, they never terminate (see p. 142). Rational numbers can be written as quotients p/q of integers p and q; this is not possible for irrational numbers. A solution can be approximated, however, in the form of a summation of quotients.

The approximation to the irrational number of the golden section takes the form shown below, where w_0 is the largest whole number less than W, and the small remainder, $r_0>0$, is expressed as an infinite sum of quotients. W is therefore equal to w_0+r_0. The summation can be broken off after the nth term.

The series of quotients for the approximation of the golden mean is $W_n = 1, 1/2, 2/3, 3/5, 5/8, 8/13, 13/21\ldots$; the rule for forming Fibonacci numbers is clearly seen. The sum of the two preceding numbers – here, the denominators – gives the next denominator.

Fibonacci series:

1, 1, 2, 3, 5, 8, 13, 21, 34, 55...

The quotient formed by neighboring members approaches the value 1.618... (golden mean)

Golden section:

The longer part ($= d$) is to the whole ($= 1$) as the shorter part ($= 1-d$) is to the longer:

$$d:1 = (1-d):d$$

$$W = w_0 + \cfrac{1}{w_1 + \cfrac{1}{w_2 + \cfrac{1}{w_3 + \cfrac{1}{w_4 + \cfrac{1}{w_5 \ldots}}}}}$$

Approximation: $IW-W_nI \leqq \text{const}/q_n^2$

Of course, the golden section plays an important role in architecture and the fine arts as well. Although the golden section is often used intentionally, it is also frequently employed unconsciously. Figure 6.14 shows a painting by Seurat in which the "golden" ratios and sections have been highlighted[11].

Fig. 6.14. Golden sections in a painting by Seurat.[11]

The golden mean is the most irrational of all irrational numbers and thus has something to do with chaos. In certain trajectories and mathematical or graphical descriptions of complex dynamical systems, an increase in nonlinearity results in greater chaos. Eventually, only a few boundary curves separate the regions of chaos and, finally, these shrink to just one. This last curve can be related to the golden mean in the way discussed above. One more indication of harmony on the boundary between order and chaos?

The most irrational trajectories, that is, those based on the numerical ratio of the golden mean, are the most likely to survive perturbation. They resist the outbreak of chaos the longest[6].

Does taking a bull by the horns give rise to beauty? Does beauty emerge when a dynamical system just barely succeeds in avoiding chaos? Is beauty like hiking along a mountain ridge?

> Henceforth I know why rosebud pleases
> Now as the time of roses ceases
> A late bloom still in solitude
> Replaces Nature's multitude.
>
> J. W. v. Goethe

Is the "dying rose" a symbol of supreme beauty only? Might it not also herald an objective theory of aesthetics? * Viewed in this way, beauty could be understood not

* I thank Wolfgang Kaempfer, Triest, for contributing important insights to this train of thought.

only as subjective perception, but also as perception based on an underlying, mathematical law defining the boundary between order and chaos.

Is "beauty," then, not only a question of cultural prescription and social convention, but also an intrinsic property of things and the world? Does the world at the boundary between order and chaos have an essentially harmonic structure? Let us listen to the simple words of a poet:

> She blooms because she blooms,
> the rose...
> Does not ask why
> nor does she preen herself
> to catch my eye [12]
>
> Angelus Silesius

Fragile Beauty – A New Artistic Norm

The fractals discussed and shown in this chapter give rise to figures of great aesthetic appeal. Their beauty is undeniable. As already noted, this aesthetic category is revealed precisely in the regions of transition between order and chaos, in the fractal regions. I would like to present several more examples from outside the physical sciences.

Albrecht Dürer's self-portrait, which hangs in the Alte Pinakothek in Munich, is well-known: it shows a handsome young man with flowing curls and an immaculate moustache, an ideal of beauty. I have allowed myself to divide this portrait down the middle and to put the two parts back together in such a way that two faces are formed, one from the two left halves, the other from the two right halves. Strikingly, the two pictures are totally different; not only Dürer's face but all human faces are unsymmetrical. Furthermore, the two artificial portraits are utterly lacking in appeal (Fig. 6.15). Every tension has vanished from the face, and our interest in it wanes. Clearly, art is not total harmony; it is not perfection. This was known not only to Western artists but to classical Islamic artists as well. They introduced intentional asymmetries into elegant, highly stylized works of art – because pure symmetry was the province of Allah alone!

Beauty appears to be most captivating and transparent wherever it verges on the border of chaos, where its order is put in peril. Beauty is a narrow path between two precipices; on the one side, dissolution of all order in chaos, on the other, a frozen world of symmetry and order. Only along this perilous path does beauty take on form. I would like to illustrate this through three examples.

Fig. 6.15. The self-portrait of Albrecht Dürer in the Alte Pinakothek in Münich. Left: Normal reproduction. Top right: A face formed from the two right halves. Bottom right: A face formed from the two left halves.

El Greco, who had devoted himself entirely to the stagnating art of painting Byzantine icons before fleeing Crete after the Turkish conquest, burst the bounds of form in unforeseeable ways in his new country, Spain. I am thinking of the picture "Thunderstorm over Toledo," which hangs in the Metropolitan Museum in New York. An explosion of color and form, just barely holding back from the brink of chaos. It is no wonder that El Greco had difficulties with the Inquisition.

A second example is Wassily Kandinsky. I am thinking of the period between 1908 and 1910, which is documented so well in the Lenbach House in Munich. Before this time, Kandinsky had painted in a conventional, decorative, aesthetically appealing Jugendstil manner. In 1908 he began to abandon the object-oriented style, but continued to paint recognizable objects. But his paintings gradually became more and more complex with unbelievable explosions of color. By 1910/11 this fantastic phase of transition was complete. Objects had been completely abandoned and an abstract, constructivist style of painting began to emerge. It is impossible to say whether this style contains chaotic elements or instead represents a new form of order, threatened by stagnation.

To create art on the border between order and chaos sometimes represents too great a challenge and peril to both artists and their works. Friedrich Hölderlin ventured too close to this beguiling and dangerous border, from which he then retreated into mental illness.[13] I would like to illustrate this through his poem "In lovely blueness," which, as stated in histories of literature, originated and was written during a period of spiritual darkness after Hölderlin had already spent some 15 years in the Tübingen tower on the Neckar river. The poem begins in a simple and straightforward way, well-controlled, orderly. All at once in the middle, the fire bursts out, a threatening, unbearable chaos is revealed. Then, just as suddenly, it vanishes. This is the only eruption in the poem, which then reverts to its tranquil, orderly form.

> In lovely blueness with its metal roof the steeple blossoms.
> Around it the crying of swallows hovers, most
> moving blueness surrounds it. The sun hangs
> high above it and colours the sheets of tin, but
> up above in the wind silently crows the
> weathercock...

And the poem continues on tranquilly for quite some time. Suddenly, it erupts:

> Is there a measure on earth? There is none.
> For never the Creator's worlds constrict the
> progress of thunder. A flower too is beautiful,
> because it blooms under the sun. Often in life
> the eye discovers beings that could be called
> much more beautiful still than flowers. Oh,
> well I know it! For to bleed both in body and
> heart, and wholly to be no more, does that
> please God?

Here, the unbearable hike along the narrow mountain path ends. The descent, indeed the fall, begins. Order verging on the abstract becomes more and more apparent. A kind of mental exhaustion is manifested.

> Yet the soul, it is my belief, must remain pure,
> else on pinions the eagle reaches far as the
> Mighty with songs of praise and the voice of so
> many birds. It is the essence, the form it is.
> You beautiful little stream, you seem touching,
> as you flow so clear, clear as the eye of
> divinity, through the Milky Way.[14]

Palindromes – Islands of Order in the Genetic Script

"Hidden harmonies are more powerful than revealed," said Heraclitus. Let us take another look at the genetic script. It consists of a four-character code, which is used to produce a linear set of instructions specifying the structure of an organism (see chapter 3). Does the DNA contain this information only or are there higher structural elements, superstructures, that help to catalog the library of information? Indeed, structures of this kind are discernible at certain sites. They are palindromic in nature. A palindrome (Greek *palindrom,* to go back and forth) is writing that, read forward or backward, gives the same message. Typical palindromes are the word madam, the year 1991, and the following sentences: "Madam, I'm Adam"; "A man, a plan, a canal. Panama."

Palindromes as superstructures are necessarily very artful fabrications, since, despite the imposed conditions (readability both forward and backward), the sense of the underlying writing must not be lost. They verge on the boundary between order and chaos. Consequently, palindromic sentences are often quite droll.

Palindromic sequences are in fact present in the double helix. They apparently serve there as some sort of higher-level recognition signal along the relatively monotonous information band of DNA. In DNA, the palindromes are arranged on opposing strands in such a way that the same order of bases in the 5′→3′ reading direction is established and the base-pairing rules are fulfilled – difficult and demanding conditions! Three DNA palindromes are shown in Figure 6.16; they represent cutting

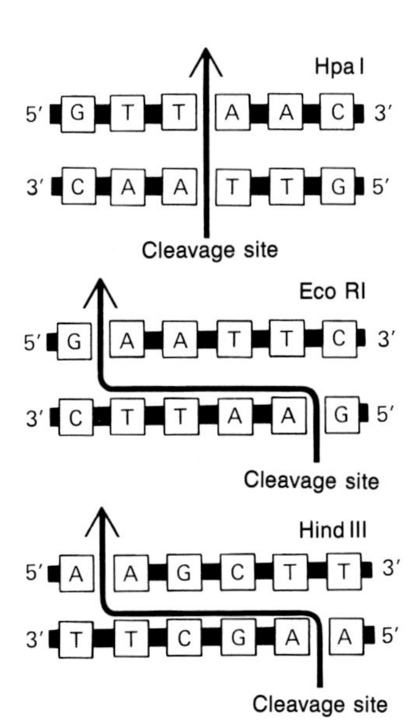

Fig. 6.16. Nucleotide sequences recognized by three commonly used enzymes that cut nucleic acids into pieces (restriction enzymes, p. 57). Such sequences are often six base pairs long and palindromic.

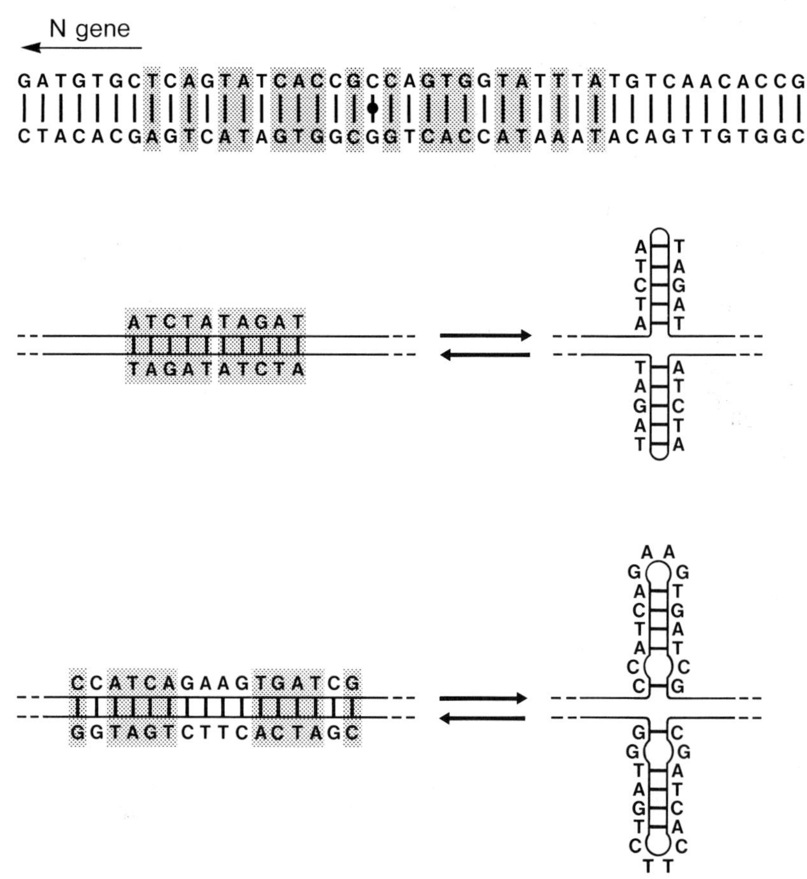

Fig. 6.17. A 45-base-pair-long piece of λ DNA from the region of the operator gene O. The (incomplete) palindrome is symmetrically arranged around the point shown. The middle and bottom parts show how palindromic sequences protrude as hairpin loops, which preserve the number of base pairs.

sites for the restriction enzymes mentioned in chapter 3 (p. 57). The corresponding restriction enzymes exclusively recognize these palindromic sequences, read on the upper strand from left to right and on the lower from right to left. These enzymes, which are of central importance in genetic engineering, thus cleave the nucleic acids at palindromic sites. But palindromes also play an important role in the normal regulation of the genetic instructions by so-called repressors. The genetic information is not called up from all the genes at the same time and to the same extent. Were this so, sheer bedlam would reign in the cell. Instead, most genes are blocked by repressors, which bind firmly at certain sites – namely, the respective palindromic sequences – and act like bicycle locks to prevent the genetic wheel from turning. Only

Fig. 6.18. A spring in the courtyard of the monastery Moni Preveli on the south coast of Crete.

specific molecules acting as keys are capable of removing these locks and only then is the genetic information at this site read off.

How do blocking proteins or restriction enzymes recognize these highly symmetrical sites? In principle, there are two possibilities. First, owing to their intrinsic symmetry, palindromic sequences can form base-paired loops (hairpins, cf. Fig. 6.17). These precisely defined loops protrude from the otherwise featureless stretch of DNA and could serve as recognizable structures. The role of DNA loops and the influence of neighboring regions on biologically important structural changes in DNA have been demonstrated only recently.[15]

Second, palindromic symmetry could also serve in a simple way for recognition. Since binding proteins are likewise symmetric, they could exploit their symmetry to bind to the symmetrical DNA. Although it is still not possible to make a precise distinction between these two mechanisms, such palindromes clearly represent islands of “high-level order” surrounded by the informational script of DNA.

One of the most beautiful palindromes of a nongenetic kind, one involving language, I found at a spring in the courtyard of a monastery on Crete. It is shown in Figure 6.18. This artfully fabricated sentence makes good sense and just at the place where it is engraved:

Wash away your sins, do not wash your countenance only.

Paul Celan

Psalm [16]

No one moulds us again out of earth and clay.
no one conjures our dust.
No one.

Praised be your name, no one.
For your sake
we shall flower.
Towards
you

A nothing
we were, are, shall
remain, flowering:
the nothing-, the
no one's rose.

With
our pistil soul-bright,
with our stamen heaven-ravaged
our corolla red
with the crimson word which we sang
over, O over
the thorn.

7. Big Bang – Idea or Matter?

A dialogue between Werner Heisenberg and Wolfgang Pauli on physics, metaphysics, and religion[1]

PAULI: *Are you of the opinion, then, not only that physics consists of experiments and measurements on the one side and a formal mathematical framework on the other, but also that true philosophy ought to be pursued at the boundary where the two sides meet? This means that, using natural language, one should try to explain what actually occurs in this interplay of experiment and mathematics. I suspect that all the difficulties in understanding quantum theory arise at just this boundary, which the positivists* * *usually pass over in silence for the simple reason that precise concepts cannot be employed there. The experimental physicist must be able to discuss his experiments and is thus compelled de facto to employ the concepts of classical physics, which, as we now know, do not correspond exactly to nature. This is the fundamental dilemma; it cannot simply be ignored.*

HEISENBERG: *The positivists are indeed extremely sensitive to all questions possessing, as they say, a prescientific character. I am reminded of a book on the law of causality in which certain questions or formulations are repeatedly dismissed because it is alleged that they involve metaphysical relics from a prescientific or animistic epoch of thinking. For instance, the biological concepts of "holism" and "entelechy" are rejected as prescientific and an attempt is made to prove that statements in which these concepts are commonly used cannot be tested. The word "metaphysics" is regarded as a kind of swearword to label utterly vague trains of thought...*

It would be preposterous, it seems to me, if I were to try to forbid the questions or trains of thought of the early philosophers because they were not expressed in precise language. Naturally, I often have difficulty in understanding where these trains of thought are supposed to lead. In such cases, I try to translate them into modern

* Positivism is an antimethaphysical philosophy, which accepts only the authority of positive experience.

terms and then determine whether we are able to provide new answers. But I am not at all reluctant to reconsider the old questions, just as I am not reluctant to employ the traditional language of one of the old religions. We know that religion uses a language of metaphors and parables, which never quite convey exactly what is meant. But, in the final analysis, most of the old religions, which originated in an earlier epoch than modern science, have the same content, the same core of ideas, which needs to be conveyed in metaphors and parables and which is centered on the question of values. The positivists are perhaps justified in saying that today it is often difficult to find meaning in such parables. Yet there remains the task of uncovering this meaning, since it apparently represents a crucial part of our reality, or perhaps of expressing it in a new language if it is no longer expressible in the old.

PAULI: *If you are pondering such questions, then it is obvious that you have no use for a concept of truth based on the possibility of prediction. But what, then, is your concept of truth in science?...*

HEISENBERG: *When we see an airplane in the sky, we are able to predict with a certain degree of certainty where it will be one second later. We simply extend its flight path in a straight line; or, if we have already noticed that the airplane is describing a curve, we take the curvature into account. In most cases, we are thereby quite successful. But we still haven't understood the path. Only after we have spoken with the pilot and received his explanation about the projected flight path have we really understood that path...*

PAULI: *What in nature corresponds to the pilot's intention or flight assignment?*

HEISENBERG: *Words like "intention" and "assignment" are anthropomorphic and, in nature, are best understood metaphorically. Perhaps, however, we can make some progress with the old comparison between the astronomy of Ptolemy and the knowledge of planetary motions since Newton. In terms of a criterion of truth involving predictability, Ptolemaic astronomy was no worse than the later Newtonian. But when we compare Newton and Ptolemy today, we feel that Newton's equations of motion describe the orbits of the heavenly bodies more comprehensively and correctly – that he, so to speak, described the intention according to which nature is constructed. Or, to take an example from modern physics: When we learn that the conservation laws for energy or charge, for example, are universal in character, that they apply to all areas of physics and are due to symmetry properties in the basic laws, then we are inclined to conclude that these symmetries are key elements in the plan according to which nature was created. Here, I am perfectly well aware that the words "plan" and "created" are once again taken from the human sphere and should therefore be regarded at best as metaphors. Yet it is equally apparent that, in this case, language is unable to provide us with superhuman terms, which might help us come closer to what is meant. What else should I say about my concept of scientific truth?*

PAULI: *Naturally, the positivists could now object that you are being unclear and just batting around the issue, and they can take pride in the fact that such things cannot happen to them. But where is more truth to be found, in what is unclear or in what is clear? "In the abyss lies truth." Yet, is there an abyss, and is there truth? And does this abyss have something to do with the question of life and death?...*

HEISENBERG: *The positivists have a simple solution: The world is divisible into what one can clearly express and what one has to remain silent about. Here, then, one must remain silent. But there is no philosophy more full of nonsense than this one. For virtually nothing can be expressed clearly. If everything unclear is cast aside, then there remain only tautologies of no interest whatsoever.*

PAULI: *You said before that the language of metaphors and parables, in which the old religions express themselves, is not unfamiliar to you either...*

You implied that the different religions, with their diverse metaphors, embody the same "core of ideas." This content, as you formulated it, centers on the question of values. What did you mean by this and what does this core of ideas, to use your expression, have to do with your concept of truth?

HEISENBERG: *The question of values is, after all, the question of what we do, what we strive for, how we should behave. The question is posed by man and is thus relative to man; it is the question of what compass we should use to orient ourselves when we seek to find our way in life. This compass has been given diverse names in the different religions and philosophies of life. Chance, God's will, reason, to name just a few. The differences in the names indicate that the human groups that have so named their compass have deep underlying differences in the structure of their consciousness. By no means do I wish to downplay these differences. But it is my impression that all these formulations involve the relations of man to the central order of the world. We certainly know that, for us, reality is dependent on the structure of our consciousness; the objective sphere represents only a small part of our reality. But even where the subjective sphere is being considered, the central order is at work and denies us the right to regard the forms of this sphere as resulting from the play of chance or caprice. However, the subjective sphere, be it that of an individual or a people, can give rise to much confusion. The demons may be on the loose, so to speak, and up to mischief. Or, to express it more scientifically, partial orders that no longer fit into the central order, that are separated from it, may be at work. But, ultimately, what always wins out is the central order or, to use the terminology of antiquity, the "One," with which, in the language of religion, we enter into relation. When the question of values is raised, the commandment seems to be that we should act in accordance with this central order – simply to avoid the confusion potentially resulting from discrete partial orders. The effectiveness of the One is already revealed in the fact that we consider order good, confusion and chaos bad. A city destroyed by an atomic bomb appalls us; but we are delighted when a desert is successfully turned into a blooming landscape. In science, the central order is revealed in the fact that the metaphors ultimately employed are those like "nature has been created according to this plan." And at this point my concept of truth is connected with what is meant by the core of ideas in all religions. I find it has become much easier to think about these interrelations now that quantum theory is understood. For it allows us to formulate uniform orders over very broad areas in an abstract mathematical language; at the same time, we recognize that, when we wish to describe the effects of this order in natural language, we are compelled to use equations and to view things in complementary ways, which force us to accept paradoxes and apparent contradictions.*

PAULI: *Indeed, this conceptual model is perfectly comprehensible. But what do you mean when you say that the central order always wins out? Either this order is there or it is not. What does "win out" mean here?*

HEISENBERG: *By that I mean something utterly trite – for example, the fact that after each winter flowers bloom once again in the meadows and that after each war cities are rebuilt, in short that chaos is always transformed back into order...*

PAULI: *Do you yourself believe in a personal God? I know, of course, it is difficult to give such a question a clear meaning, but the sense of the question is certainly evident.*

HEISENBERG: *May I formulate the question differently? It would then be as follows: Is it possible for you or for someone else to confront directly the central order of things or events, which is not open to question, to come into direct relation to it, in the same way that the soul of another person does? I am intentionally using here the word "soul," the meaning of which is so hard to interpret, in order to avoid being misunderstood. If you posed the question in this way, I would answer yes...*

PAULI: *Why have you used the word "soul" here instead of simply speaking of another person?*

HEISENBERG: *Because here the word "soul" means, in effect, the central order – the center – of a being that, in its outward appearances, is extremely variable and hard to grasp...*

PAULI: *You mean then that for you the central order can be present with the same intensity as the soul of another person?*

HEISENBERG: *Perhaps...*

The Big Bang – A Real, Physical Event?

Modern physics has good reasons to assume that the universe originated in a "Big Bang" 15 to 20 billion years ago. Matter was formed from an massive packet of energy and hurled explosively outward from the center of the Big Bang. The elementary particles, the physical laws, and the associated matter all came into being within a few thousandths of a second. Or did matter evolve first and then the physical laws? This question is discussed, among others, in a fascinating monograph by Steven Weinberg[2].

There are essentially two pieces of evidence for the Big Bang. The first is the now old observation that the galaxies visible to us (clusters of stars, the Milky Way, spiral nebulae) are moving faster away from our point of observation on earth the further distant they are to begin with. This is revealed by the spectral lines of the galaxies. The material parts of the universe thus behave like fragments flying apart after an explosion. The fastest are those that have already gone the farthest. At the edge of what is visible to our telescopes there are galaxies whose light has taken billions of years to reach us. These galaxies cannot move faster than the speed of light. More precisely, their motion is at most only close to the speed of light. Otherwise, according to the theory of relativity, matter would become infinite. Anything flying away from us at the speed of light could not be observed optically, because its light would never reach us. This situation would also make no sense, since a galactic system with infinite mass cannot exist. Physical impossibility and logical meaninglessness thus coincide here. The second piece of evidence for the Big Bang is the so-called cosmic background radiation, a remnant of the Big Bang[2].

The question is still unanswered whether the universe is "open," that is, whether the Big Bang is a unique event, after which the universe continues forever to fly apart into endless space, or whether this motion will gradually come to a stop and then reverse itself, with everything collapsing to the Big Bang's center of origin. If the universe indeed pulsates, it should do this with a period of 80 billion years or so.

At the beginning of the universal expansion, at the moment of the Big Bang, there was an unimaginably dense and hot plasma of energy and matter. All physical states were characterized by fundamental simplicity and symmetry. At the moment of the Big Bang (or before, if that is meaningful) the universe was a "super-energy crystal" free of all defects and perturbations. The first infinitely small perturbation, deviation, or, more generally, the first symmetry break unleashed the cascade of the Big Bang. Cosmic and, subsequently, biological evolution are understandable in terms of a continuous series of such symmetry breaks or bifurcations as the universe expanded and cooled. Therefore, a parallel must exist between the history of the universe and its logical underlying structure. It could be expressed as follows: The universe evolves, it constantly produces something new, creates things, laws, relations, none of which were "predictable." In its basic structure the universe is creative.

Erich Jantsch has written[3]: "The underlying symmetries being sought are revealed, then, by returning to the historical origin. Accordingly, in the reverse direction, evolution – the unfolding of history – is characterized by a series of symmetry breaks. Fundamental symmetry breaks can indeed be traced not only through the physical-cosmological history of the universe but also through the history of life and of the mind in our local world. In each case symmetry breaks bring into play new dynamical possibilities of morphogenesis or the generation of forms and thereby signal an act of self-transcendence. Only through symmetry breaks does complexity become possible. The resulting world is less and less reducible to a single level of underlying principles, whose unity is only understandable abstractly in terms of a common origin. What arises is a multilayered coordinated reality."

The preceding chapters have introduced the biological, physical, and mathematical principles of evolution. Historio graphers of nature, particularly paleontologists, are capable of providing an insightful, albeit indirect, chain of evidence for the course of evolution. On this earth, primitive unicellular organisms gave rise to protozoa, plants, eukaryotes, worms, higher animals, and ultimately man; this is depicted in the phylogenetic tree of living organisms sketched by Charles Darwin in his 1859 work "The Origin of Species."[4] For the past 120 years the biological sciences – biochemistry, too – have largely focused on confirming Darwin's theory. In this chapter an attempt will be made to portray the biological way of thinking in a general historical context.

Newton and Darwin

Newton's worldview and ideas concerning living things were at total odds with those of Darwin. This is supported by a conversation said to have taken place between Newton and John Conduitt[5]: "Newton told him his belief that there was a sort of revolution of heavenly bodies. Light and vapor from the sun gather together to make secondary bodies like the moon, which continue to grow as they gather more matter and become primary planets and ultimately comets, which in turn fall into the sun to replenish its matter. He thought that the great comet of 1680, after five or six orbits, would fall into the sun, increasing its heat so much that life on earth would cease. Mankind was of recent date, he continued, and there were marks of ruin on the earth which suggested earlier cataclysms like the one he predicted. Conduitt asked how the earth could have been repeopled if life had been destroyed. It required a creator, Newton answered. Why did he not publish his conjectures as Kepler had done? 'I do not deal in conjectures,' he answered. He picked up the *Principia* and showed Conduitt hints of his belief which he had put in the discussion of comets. Why did he not own it outright? He laughed and said that he had published enough for people to know his meaning."

Accordingly, Newton believed in a static world, since creation by a creator is a one-time static act. Afterwards God stands aside and lets the wound-up clockwork of the world wind down.

Diametrically opposed to this view was Darwin's. His observations led him to a completely new perspective of nature. He rejected the prevailing static view of nature and tried to understand nature as a historical process. Darwin thereby established a far-reaching theory and created a new paradigm, to use a term of Thomas Kuhn[6].

What were the consequences? First of all, the paradigm was "new" in the sense that it did not seem to fit into established ideas and ways of thinking, that is, into the spiritual, technical, and economic landscape of the 19th century, which, in many respects, was still the Newtonian age. Although Newton had been dead for more than one hundred years, the consequence of his view of nature – the age of technology and machines – had just begun to make itself felt and the general ideas of the human beings employing these machines were still gradually adjusting to this new situation. The result was something approaching our modern, scientifically based view of the world, the principal ideas of which center on causality, the ability to describe everything in mathematical terms, reversibility, machines, and technology. All processes were thought to be experimentally repeatable – for example, the oscillation of a pendulum and the trajectory of a thrown stone. Life was viewed as a clockwork, which is wound up at the beginning and then gradually winds down; it was only a matter of time before the process of rewinding would be understood. This automatistic point of view dominated the idea of evolution and, in general, the science of life, even though completely different ways of thinking had existed even before Darwin, two examples being the ideas of Lamarck and Goethe (see chapter 3). The Darwinian theory came to be understood as a scheme of causality from the Big Bang to the emergence of Homo sapiens and is still viewed this way today. For this reason it is essential to reexamine the core of Darwin's theory, to read his original writings and not the secondary literature, which is full of false interpretations. The Darwinian theory of evolution offers a plausible and self-consistent explanation of the diversity of species and of life, but it is not a theory of causality. Evolution could have proceeded differently. Moreover, the theory provides no predictions for the future.

The increasing influence of the scientific approach on modern thought is reflected in the gradual disappearance of teleology from our view of the world, as Reinhard Löw[7] has documented in detail. Teleology is the doctrine that a process has a purpose and is directed to an end. When modern Darwinism transformed biology into an objective science, it banished teleology. Furthermore, Darwin's theory dethroned man as the highest creature, the teleological goal of evolution, "creation's crowning glory." Man was reduced to a mere coincidental link in the chain of evolution. Subjectively, this poses a problem for the man of thought and feeling. But *to question objectively* the theory of evolution is, of course, no longer possible. What is possible, though, is to ponder whether life as an integrated phenomenon is indeed accessible to exhaustive scientific analysis by our objective sciences. The theory of evolution is a comprehensive theory of nature possessing axiomatic character; it is neither provable nor disprovable. That is, although the theory is capable of accounting for many facts in an insightful way, it is open to neither mathematical nor physical proof. It does not solve a puzzle once and for all. (Wittgenstein: "There is always a solution to a puzzle."[8])

Darwin established a new paradigm for a new view of nature, but it cannot be derived or proved, nor indeed is this necessary. It has led to a comprehensive and unified view of living things – it is plausible, but, as a "worldview", it is not provable.

Preoccupied as we are with the daily challenges of "problem-solving research," we all too readily forget the paradigmatic nature of science. Scientific thought and action occurs within a system of axioms, basic assumptions, paradigms, which as such remain unquestioned in a particular branch of science or during a certain scientific epoch. In a "scientific revolution," however, this system of paradigms can be overthrown[6]. Prior to Darwin, the living world was static; afterwards it was evolving. Prior to Max Planck, energy could always be divided into smaller and smaller packets; afterwards it was carried only in discrete quanta. Yet no scientific description is a "complete" description (see chapter 9). This realization compels us to ask certain questions: What concept of life is open to us and how encompassing is the system of life set forth by Darwin? The concept of life is certainly not that of Goethe and also not what is meant when we talk of feeling alive. John Haldane, a biologist who definitely thinks in material terms, had the following to say (cited in Reinhard Löw[7]): "For the biologist teleology is like a mistress: he cannot live without her, but he does not want to be seen with her in public."

But life is not merely "the mode of being of proteins."[9] Darwin applied his mental acuity to the examination of materials he had collected during his travels around the world and gradually developed a general theory of the origin of species. And a whole science has proven him right, not only the early phenomenological investigations of biology and zoology, but also modern molecular biology. By deciphering the genetic code, molecular biologists have gone back to the "source" itself and are now able to determine exactly where certain parts of the text of life were copied and how often this copying introduced changes. Indeed, their methods of research are similar to those of philologists searching through handwritten medieval manuscripts in dusty libraries. Paleontologists, too, are now able to date their finds by various methods and thereby provide a nearly unbroken chain of evidence in support of Darwin's theory. In short, the static view of nature has been banished forever and replaced by a dynamical one.

Darwin's dynamical view of natural history has also led to some seemingly unpleasant consequences. No longer does life offer us a firm grip; everything flows. About 5 million species exist in the world at present. But roughly 500 million species have emerged throughout the history of the earth. In other words, what we now see represents only one percent of what has existed at one time or other. Perhaps man, too, is but a transitional stage[10]. Evolution devours its children[11].

The Complexity of Living Things

What are the consequences of all this for our way of thinking? Predictability is no longer a criterion of scientific truth. In Newtonian systems, the trajectories of projectiles and the orbits of the planets can be calculated and predicted given a set of initial conditions. For this very reason the famous discovery of the planet Neptune in 1846 was celebrated as the keystone in establishing the validity of the materialistic-deterministic worldview[9].

Ever since the introduction of quantum physics and the theory of relativity, the situation in the natural sciences has been undergoing a gradual but profound change. We have now reached a limit in our ability to describe living things. This situation is in some ways analogous to that brought about by the Heisenberg uncertainty principle in the description of *elementary particles.* In systems with bifurcation points predictability is limited[12] (see chapter 4, p. 109).

Saying farewell to the predictability of many physical events in the macroscopic world, too, implies by no means that science has reached its end, that we must now begin to "...admire quietly the unfathomable" (Goethe). It means no more and no less than that man must say farewell to the "myth of predictability." The Newtonian way of thinking in terms of the general linearity of differential equations represents an unacceptable simplification in the modern scientific treatment of fundamentally complex systems. The description of such systems requires a new theory of transformation, like the baker transformation (see chapter 5). The phylogenetic tree is representable only by a transformation that allows for unpredictable branch points. Something similar holds true for the dynamical functions of the central nervous system, a network of hierarchically connected decision-making processes involving feedback. The first theoretical approaches have already been made in this direction. The complexity of living things sets a limit to what we are capable of knowing about them. Not that we are unable to describe, say, many of the details of nucleic acids and proteins. But the interaction of these components in subsystems and higher organizations results in an unpredictable network system, which possesses the character of fundamental complexity[13].

Clearly, one should avoid interpreting all this as a possible approach to the proof of God's existence by concluding, say, that God is able to intervene at the bifurcation points. Kant has already demonstrated the impossibility of providing a scientific proof of God's existence. Moreover, it has always proven fatal to exploit "the good Lord" as a stopgap wherever our knowledge fails us. It is easy to overestimate science and to overlook the prerequisites of scientific knowledge. The only way to pursue science is according to the method of Descartes, who, as already quoted in chapter 1, said in his discourse "De la mèthode"[14]: "If a problem is too complex to be solved at once, then break it up into problems that are small enough to be solved separately."

This statement implicitly assumes, of course, that after all the separate problems have been solved, they can be put back together, like mosaic tiles, to produce an overall picture. In this way, a single solution is obtained for the whole complex problem.

In multiparameter systems subject to a high degree of feedback, however, this assumption or, to be more precise, this hope proves to be fundamentally wrong. These systems are not reducible. I refer to such systems as fundamentally complex; they possess the property that the whole is more than the sum of the parts. There is no reversibility in these systems. Irreversible thermodynamics has to be applied instead of classical, reversible thermodynamics. For this reason it would simply be intellectually negligent to assume that, in sciences such as biochemistry and neurophysiology, the mosaic tiles can be fit back together to obtain the complete picture of a living thing.

Self-organization

In this section I will try to examine more closely the overused concept of self-organization and to shed some light on its actual meaning. To do this, I would like to distinguish the different levels of self-organization.

Self-organization through Inherent Properties

In the simplest case, the external form of macroscopic bodies is governed by the "packing" of their component molecules or atoms. Crystals, for instance, consist of numerous particles of the same kind. Just as marbles arrange themselves in a box into regular honeycomb-like structures (hexagonal close packing) and cubic building blocks fit together compactly, so, too, do the regular arrangements in crystals arise. Here, then, the self-organization is governed by the prespecified properties of the identical components. The base pairs in a double helix likewise represent a form of self-organization based on a prespecified molecular structure. Lipid layers, soap bubbles, dewdrops, and frostwork are also easily explained in terms of the self-organization of their underlying components.

Self-organization in Ontogenesis

Ontogenesis refers to the formation of an organism starting from the undifferentiated egg cell and proceeding to the highly differentiated organism in its entirety with internal organs, external limbs, sense organs, etc. Apparently, there is an inherent "tendency" for the initially unordered mass of cells to take on a certain form. This is difficult to grasp. In fact, it is tempting to believe that magical interactions or long-range effects are at work. For this reason the wonders of nature have seduced many people to believe in the notion of a vital force, the action of a world soul, God's hand, entelechies (from the Greek εν τελοζ εχειν); to have an end in itself; according to Aristotle, the inherent, directing force in matter). But an attempt should always be made to trace natural phenomena to physical forces and material structures. Here, too.

Using a simple organism, a small freshwater polyp called hydra, Alfred Gierer was able to show that head formation is dependent on a specific growth factor, the concentration of which governs the development of the head[15,16]. This polarization of the organism is most likely due to two competing effects involving feedback: activation of growth on one side and inhibition on the other. In chapters 5 and 6 we have already learned that such systems generate order. The forms of blossoms and the symmetries of plants are explainable in this way. The same principle always holds: in dynamical systems, two opposing principles interact and lead to the dynamical creation of order. Order arises through the formation and decay of structures.

The formation of structures through "self-organization" is observed, for example, when a freshwater polyp is cut into two pieces or – as shown schematically in Figure 7.1a and b – a piece is snipped out of the middle. Within 48 hours a new head forms in one part of the initially rather uniform tissue (d). The first event in the formation of the head is the generation of a "morphogenetic field." Within a few hours the eventual head region is "activated" – indicated in the schematic drawing (c) by hatching – and this, at first invisible, activation is responsible for the subsequent formation of the new head.

Evidence for the initial generation of a morphogenetic field is not easy to obtain, however. It requires experiments with tissue grafts. The principle is illustrated schematically in Figure 7.1b and c. If, immediately after regeneration has begun, a part of the eventual head region is transplanted into tissue not too far away from the head region of another hydra, usually the grafted tissue does not cause any change to occur; instead, it remains part of the gut region (b) – the eventual head region is not yet activated. If this operation is performed about six hours after regeneration has begun, however, the grafted tissue induces head formation in the gut region (c). After six hours, then, the regenerating tissue has already been activated to form the eventual head region. Activation probably means that there is a high local concentration of an activating substance. The formation of the morphogenetic field, then, is equivalent to the generation (Fig. 7.1e and f) of a concentration gradient of a growth factor (activator), starting from a uniform distribution of the activator (e) and leading to a higher concentration at one end of the tissue (f). The polarity of the organism is thus governed by a field, referred to as a "morphogenetic field." What is the biochemical structure of this field?

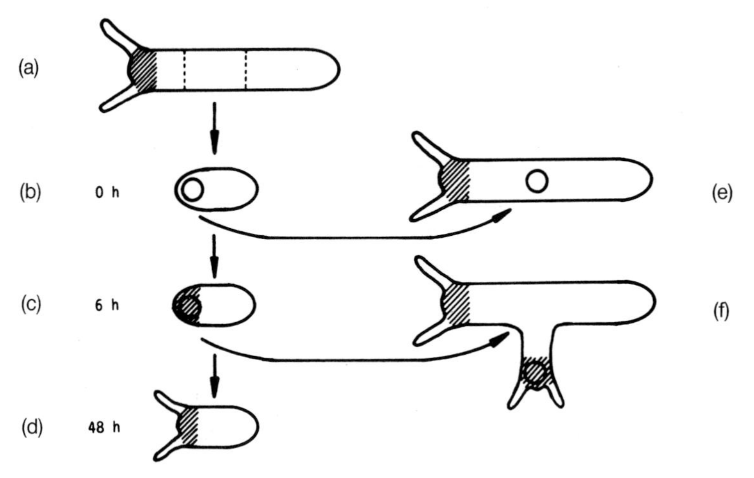

Fig. 7.1. Structure formation or morphogenesis by self-organization in the "morphogenetic field"[15] as exemplified by hydra, a small water polyp.

In this case, the "field" simply refers to a concentration gradient of a certain activating substance, formed at the eventual head region and actively degraded and thereby inactivated by a counteracting substance at the other end of the organism. Once again, we encounter a dual, morphogenetic principle involving feedback. The term "field," in this case, is not really correct; it would be better to speak of a concentration gradient.

The situation is more complicated in higher organisms. Here, there are apparently higher-level "control genes," which code for entire groups of signals or signal substances. For instance, certain mutations of drosophila, a small fly, cause legs to grow in place of antennas (Fig. 7.2). These so-called homeotic mutations are of great importance for understanding morphogenesis. Using this approach, the group of Nüsslein-Vollhardt at the Max Planck Institute for Developmental Biology in Tübingen has obtained highly interesting results[17].

Morphogenesis or the generation of forms in "morphogenetic fields" is thus explained by concentration gradients of activating or inhibiting substances whose exact nature is still unknown, but whose production appears to be governed by genes. Otherwise, these forms would not be mutable or inheritable. What is involved, then, is not self-organization in the true sense, but rather organization according to a prespecified program. This program is laid down in the DNA, possibly in a somewhat more complicated form than is the case for simple structural genes. Here, a higher control gene switches entire groups of structural genes on and off. In principle, however, this is no different from turning single genes on and off. The structure of the organism, then, is organized according to a program. But what organizes this program?

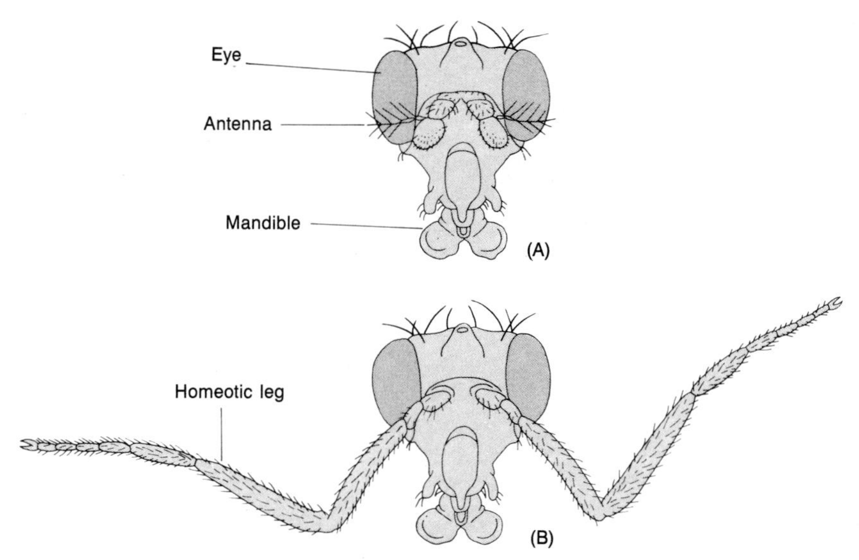

Fig. 7.2. The head of a normal adult drosophila (A) in comparison to that of a fly with a homeotic mutation of the antennapedia (B). The fly shown here exhibits an extreme form of the mutation. Usually, only parts of the antennas are transformed into leg structures[18].

True Self-organization

True self-organization is neither inherent in the physical nature of the building blocks alone nor "preorganized" by some sort of program. Naturally, structures cannot form in violation of physically imposed constraints. A soap solution can produce firm white shaving cream or delicate iridescent soap bubbles, but not cotton or hummingbirds. True self-organization is instead a property of the entire system. Under precisely defined conditions, a system with a high degree of complexity organizes itself. In the preceding chapters, we have encountered numerous examples of such systems. The scheme of self-organization relevant to biology is the Eigen hypercycle discussed in chapter 4.

Without going into a detailed discussion of Einstein's question "What for?" (see chapter 4), we still need to pose the following question: To what extent does the phenomenon of self-organization exist? When and on what substrates is it active?

Self-organization as a Physical Principle

According to what has been said so far, I think that the self-organization of matter to produce life may be understood as a physical principle. Since the Big Bang, self-organization has been a physical attribute of matter, just as mass is a physical attribute of matter and electricity a physical attribute of electrons. Why and for what purpose matter has been endowed with these physical principles and attributes are questions that cannot, and indeed should not, be accessible to scientific inquiry. The answers are beyond the realm of science, which owing its very nature cannot address questions of this kind. The natural laws are above all empirical. Matter is heavy and falls; accordingly, Galileo sought the underlying laws of falling bodies. Once they were found, attempts were made to relate them to similar laws – for example, the laws of planetary motion – in order to derive a higher, more general theory. Newton did just that with his gravitational theory. The question why gravity exists, however, is not a physical but, at best, a philosophical question, perhaps even a question of faith.

Electricity flows. Accordingly, attempts were undertaken to identify the basic particles making up this flow. The electron was thereby discovered; it has a single negative charge and its rest mass is 1/2000 that of the hydrogen atom. The question why electrical forces exist at all is a metaphysical question. Equally so are the following questions: Why is there self-organization?; Why is there life? Philosophical and religious answers have been given by the doctrines of vitalism and dialectical materialism, for example. But none of these approaches provides scientific answers. All we can do is to explore the corresponding empirical laws; that is, the laws of evolution, the Eigen hypercycle (chapter 4), the Verhulst law (chapter 6), Mendel's laws, the law of entropy. Then we seek to formulate the more general theory that matter possesses the fundamental property of self-organization. As long as the simple explanation that self-organization is a property of matter suffices, it would make little sense to seek a more complicated explanation – say, that matter is dialectical or that entelechies act in living things or that long-range, mysterious spiritual forces are at work in life.

The assertion that self-organization is a fundamental property of matter means, at the same time, that this matter is a priori filled with ideas. It carries within it the *idea* of its self-organization, its self-realization, all its blueprints and physical manifestations. Accordingly, the idea of human consciousness must have existed as a possibility at the very moment of the Big Bang. From this point of view there is no opposition between spirit and matter. In any case, spirit cannot have arisen as a superstructure from matter. The opposite is more likely: a matter devoid of ideas, without the idea of its self-organization, does not exist, any more than weightless matter exists. (The expression "weightlessness" in, for example, space travel is physically meaningless. In a space station, objects and humans are only "weightless" because the earth's gravitational attraction and the centrifugal force due to its orbital motion exactly balance.) But ideas (in the Platonic sense) may very well exist without matter. To manifest themselves (something quite different from placid existence), however, they require the presence of matter. Does this imply a return to vitalism? The central tenet of vitalism is the inherent presence in living matter – in fundamental contrast to nonliving matter – of a vital force that "knows" what it wants. In a certain sense, it has its goal in sight: the fully developed living thing or Homo sapiens as evolution's crowning glory.

I do not think so! The theory proposed here, that self-organization is the basic property of matter (just as heaviness is), is much better suited to resolve the antithesis between "dead" and "living" matter. It renders equally obsolete both the pure materialism of La Mettrie and vitalism. Moreover, it is consistent with our knowledge of biochemistry, evolution, and ontogenesis. One consequence, of course, is that the rigid concept of matter prevailing so far is no longer valid. This point will be discussed in more detail below.

The Evolutionary Field

A comprehensive and acceptable understanding of how biological forms and shapes arise presupposes a knowledge of physical biological structures. During the last two decades these have been studied so intensively that we now understand at least their principles, even if every detail is not yet known. Furthermore, an understanding of biological structures requires a knowledge of the physical and mathematical laws governing the formation of structures. These laws have now become the subject of increasing study (see chapters 4, 5, and 6). However, the generation of forms will ultimately be explainable in neither structural nor mathematical terms alone. The opposition between "material" thinking and "mathematical" thinking is as old as philosophy itself. The name Democritus is associated with thinking in terms of rigid matter, the name Pythagoras with thinking in terms of mathematical concepts and symmetries.

Are the theories at hand sufficient for a complete understanding of the development of living things? The theory coming closest to providing a mechanism is Eigen's theory of hypercycles. Viewed from a historical perspective, the Eigen hypercycle corresponds to the Galilean and Newtonian laws of motion; Newton's laws afford a general mathematical description of any motion, Eigen's hypercycles offer a framework to describe any process of biological evolution. To unify the laws of motion, Newton invented the concept of a gravitational field, which provides an answer to the question of when and where matter is heavy. For self-organization, we must ask a similar question: To what extent, when, and where does matter evolve? So far, matter has been exhaustively defined in terms of mass (weight) and inertia. Einstein relativized mass by relating it to energy through the equation $E = mc^2$. Mass becomes measurable in a gravitational field. Newton postulated this "field" in order to explain numerous mechanical phenomena (trajectories, the motion of falling bodies, oscillations, the motions of heavenly bodies) in a uniform way. The introduction of the (admittedly invisible, unreal, even paradoxical) concept of a field made it possible to explain diverse natural phenomena.

In physics, a field is defined as the entire set of possible values of a physical quantity, assigned to points in space; a material object to which these quantities are assigned need not be present. Newton introduced his field theory in 1686 to summarize Kepler's mechanics of heavenly bodies, Galileo's terrestrial mechanics, and his own laws in one unified theory. This concept of a field, to which we have now grown accustomed, was by no means unproblematic when it was first introduced. Newton, who had dealt a lot with astrology – which, of course, involves "action at a distance" – apparently borrowed the term from there, but then gave it an exact mathematical definition. This did not go unopposed by his contemporaries, notably Leibniz, who accused him of introducing mysterious long-range actions into physics. In the meantime, though, the concept of a field has come to be employed throughout physics as the foundation of virtually every theory and is no longer questioned.

To understand and unify living systems, I would like to propose the term "evolutionary field," which encompasses all events and earlier physical explanations (Big Bang, generation of forms, chaos-order relationships, hypercycles). Evolution takes place in three-dimensional space and in time. Its events are irreversible because of the directed nature of time. The scheme on page 175 summarizes my theory of the evolutionary field and compares it to gravitational theory.

"Self-organization" is a drastically shortened expression for a basic property of matter: "self-organization (the generation of forms) in the evolutionary field." Accordingly, self-organization is not merely the accretion of matter, but rather an inherent property and attribute of material substances. Self-organization is the creative potential of evolving matter – and this holds for all matter.

My theory of the evolutionary field is not without its predecessors. The British biologist Rupert Sheldrake has presented a theory of the morphogenetic field[19]. Unfortunately, Sheldrake does not always tread with due caution when he approaches the narrow boundary separating physics from metaphysics. Natural scientists are permitted to roam only within the fields of physics, chemistry, biology, etc. Certainly, they are also welcome to philosophize. But when doing so, they must clearly indicate this. In my book, I have striven to keep to this path. Sheldrake misses many good approaches by rambling along the boundary between physics and metaphysics. His "morphogenetic field" is nothing more than a rehashing of Schelling's philosophy of nature mixed with elements of vitalism and mysteriously interpreted results of modern biology.

A different, still not thoroughly formulated, proposal is due to René Thom[20]. Thom says, among other things: "A theory of morphogenesis must, of course, be applicable to biological morphogenesis; in fact, I was led to develop the qualitative dynamics presented here through reading works on embryology (particularly the books of C. H. Waddington), and my aim was to give mathematical sense to the concepts

Comparison of natural laws and theories in the gravitational field (left) and in the evolutionary field (right)

General experience	
Matter is heavy, sluggish	Matter organizes itself, forms patterns
Early attempts at description	
Aristotle: Weight is the number of Democritean atoms	*Aristotle*: Entelechies *Thomas*: Self-organization is God's organization
Empirical natural laws	
Galileo: Laws of falling bodies laws of the pendulum *Kepler*: Planetary motions *Newton*: Laws of motion	Entropy law Evolution of stars *Mendel*: Laws of heredity *Verhulst*: Growth Radioactive decay Natural clocks *Eigen*: Hypercycles
Theories	
Newton: Gravitational field	*Cramer*: Evolutionary field
Summary	
There is a gravitational field in which matter is heavy. Heaviness and the gravitational field are inseparable from matter. The gravitational field exists in three-dimensional space	There is an evolutionary field in which matter is organized. Self-organization and the evolutionary field are inseparable from matter. The evolutionary field has irreversible time as the fourth dimension

of the *morphogenetic field* of embryologists and the *chreod* of Waddington. I know very well how these concepts are at present denigrated by biologists, who criticize them for giving no chemical explanation of epigenesis. I think, however, that from an epistemological point of view an exclusively geometrical attack on the problem of morphogenesis is not only defensible but perhaps even necessary. To declare that a living thing is a global structure is merely to state an obvious fact and is not to adopt a vitalist philosophy; what is inadmissible and redolent of vitalist metaphysics is to explain local phenomena by the global structure. Therefore the biologist must, from the beginning, postulate the existence of a local determinism to account for all partial microphenomena within the living being, and then attempt to integrate all the local determinisms into a coherent, stable global structure. From this point of view the fundamental problem of biology is a topological one, for topology is precisely the mathematical discipline dealing with the passage from the local to the global.

Pushing this thesis to the extreme, we might look upon all living phenomena as manifestations of a geometric object, the *life field* (*champ vital*), similar to the gravitational or electromagnetic field; living beings would then be particles or structurally stable singularities of this field, and the phenomena of symbiosis, of predation, of parasitism, of sexuality, and so forth would be interactions and couplings between these particles. The first task is, then, the geometrical description of this field, the determination of its formal properties and its laws of evolution, while the question of the ultimate nature of the field – whether it can be explained in terms of known fields of inert matter – is really a metaphysical one. Physics has made no progress in the understanding of the ultimate nature of the gravitational field since Newton; why demand, a priori, that the biologist should be more fortunate than his colleagues, the physicist and the chemist, and arrive at an ultimate explanation of living phenomena when an equivalent ambition in the study of inert matter has been abandoned for centuries?"

Here, the great theoretician and creator of catastrophe theory remains very general, but his "champ vital" indicates that a description of the structure of living systems in terms of the currently accepted concept of matter is inadequate. Aristotle himself revealed an awareness of this state of affairs when he disagreed with the Democritean concept of matter and coined the term "entelechy" – a term that, of course, is inadequate in modern-day science. Finally, it should be mentioned that the physicist Erich Jantsch has pondered the concept of "self-organization" as well. His thoughts – directed more at the physical aspects of self-organization – are summarized in his very readable book[21].

He writes: "This new picture of science, which is primarily oriented toward models of life, rather than mechanical models, has not only brought about change in science. Thematically and with respect to the kind of knowledge, it is related to those other events that, at the beginning of the last third of our century, have signaled a metafluctuation. The basic themes are the same everywhere. They are summarized in terms like self-determination, self-organization, and self-renewal and in the recognition of a systematic interconnection of all natural dynamics throughout space and time, in the logical primacy of processes over structures, in the role of fluctuations that overrule the law of mass and give the individual and his creative insights a chance, and, finally, in the openness and creativity of an evolution that is predetermined neither in its newly forming and its vanishing structures nor in its end effect.

Science is at the point of recognizing these principles as general laws of a natural dynamics. Applied to man and his system of life, they are an expression of natural life in the deepest sense. The dualistic separation of nature and culture is thereby overcome. There is a kind of joy in reaching out, in stepping beyond natural processes – the joy of life. There is a kind of meaning in life's connection with other processes within an all-encompassing evolution – the meaning of life. We are not at the mercy of evolution – we are evolution. To the extent that science, like so many other aspects of human life, is caught up in this multilayer metafluctuation, it overcomes its alienation from man and contributes to the joy and meaning of life."

The Concept of Matter Must Be Revised

What is the true nature of this matter, which began to unfold during the Big Bang and has given rise to all forms, which is in motion as the heavenly bodies in outer space and as humans on two legs, which touches us and harms us, which we eat and excrete, which is shaped into works of arts and piles of rubbish and into the countless objects that confront us at every step? After all, object literally means that which is thrown in our way.

Since ancient times men have pondered this matter, but never before was the concept of matter subject to so much reductionism and so devoid of meaning as it is now in our everyday life. Let us take a look a what comes spontaneously to mind when we think of "matter": hard, heavy, made up of minute particles, spiritless, bought and sold, transformable by chemistry, shapeable by man or by the powers of nature, dead, composed of atoms consisting of minute elementary particles – nuclei, electrons, etc. These are the common attributes of substances. But at no point in the history of human culture was the concept of matter so reductionist as it is today.

The Greek philosophers were the first to think about matter. For Thales of Miletus the primal substance was water, for Anaximander life-sustaining air, for Heraclitus lively fire. Thales believed "that everything was full of the Gods." Anaximander discussed the nature of the universe in prose. His student Anaximenes, quoting his master, wrote: "...the principle of being is the boundless air, from which everything is formed and from which everything that will ever form arises, even the Gods and the Divinity. All other things have their origin in what, in turn, is derived from air."

From Heraclitus we have the words: "This order did none of gods or men make, but it always was and is and shall be: an everliving fire, kindling in measures and going out in measureses." He stated further: "All things are an equal exchange for fire and fire for all things, as goods are for gold and gold for goods." If the word fire is replaced by energy, then one comes astonishingly close to the concept of matter in modern physics. Empedocles believed that the world was formed from and consists of the four elements earth, air, water, and fire, intermingling through forces of attraction and repulsion (love and quarrel). Our present concept of matter, as I have presented it above, is due largely to Democritus, naturally in modified form. He first proposed an atomic theory of matter; the word "atom" (Greek *atomos*: indivisible) originated with him. Democritus understood atoms to be the smallest indivisible units of matter. That atoms split apart and are further divisible changes nothing in the intellectual concept of the atom as the smallest basic unit of matter. All properties of matter are explainable in terms of the form, size, and position of minute particles, which are noncompressible, impenetrable, too small to be seen and hence invisible, unchangeable – namely, the atoms moving in a vacuum. Democritus explained the different qualities of materials in terms of larger or smaller amounts of atoms in a volume of space. Matter and motion are eternal. Becoming and decay represent rearrangements of the atoms; even the soul consists of atoms according to Democritus. These atoms are distributed over the entire body of a man or animal and their motion is governed by laws. The basic views of Democritus have been retained practically unchanged in chemistry, in macrophysics, and in the way we view our daily world.

A completely different line of tradition extends from Pythagoras via Plato and Aristotle to modern times. The four elements of Empedocles can be traced back to the perfect solids of Plato (compare what was said about Kepler in chapter 6). Plato differentiated between ideas on the one hand and perceived phenomena on the other. The former are unchanging and transcendent entities, the latter imperfect representations.

According to his theory, the nonempirical, ideal objects act through matter, whereby matter for its part is, in a certain sense, invested with ideas. Aristotle extended this notion further. He postulated a formless and featureless primal matter (materia prima). The materia prima is invested with creativity. It is unformed matter with the capability of self-organization. The actual existing forms of matter, the materia secunda, are formed stepwise from the primal matter, whereby its features become more and more complex. Matter necessarily passes through each individual stage. The four Empedoclean elements represent certain properties: earth is cold and dry, water is cold and moist, air is warm and moist, fire is warm and dry. The soul of living things represents an additional attribute. The development of matter is governed by individual structural features and is directed toward the evolution of certain forms. This inherent power or ability or knowledge of matter as to its direction of development was termed entelechy by Aristotle. Aristotelian philosophy was rediscovered by Albertus Magnus and Thomas Aquinas and incorporated virtually unchanged into the doctrines of Christianity. But not even the Bible defines the soul in such exact terms as Aristotelian philosophy does. Aristotelian entelechy, as adapted to Christian theology, became God's will. The scholastic concept of substance still prevails in dogmatic form in Catholic doctrine.

But western thought on matter has a third stream of tradition, which is traced directly to Plato without passing through the filter of the Middle Ages. Galileo and above all Kepler refer to Plato when they proclaim and apply mathematics as the explanatory principle of the world. This becomes very clear in modern physics. Heisenberg, too, had a Platonic concept of matter[1].

For Heisenberg and for modern physics, matter corresponds to the differing appearances of an immaterial mathematical structure. This structure is reflected in the symmetry groups and conservation laws of physical quantities. Heisenberg attempted to understand the different elementary particles as solutions of a single nonlinear field equation; its group-theoretical invariance was expected to reflect the mathematical symmetry properties of the elementary particles. Although this "world formula" must be viewed as a failure – here, what Einstein said about content and substance (see chapter 4, p. 85) is relevant – it sheds some light on the approach used in modern physics: the explanation of matter in terms of mathematical principles.

However, if matter exists in an evolutionary field, analogous to the gravitational field, then the commonly accepted concept of matter has to be revised[10]. In a certain sense, matter is now soft. It does not consist of the inert hard clumps of Democritus, but rather is receptive to the evolutionary field. It is nonlinear and therefore partly indeterminate, in accord with quantum mechanics. It is filled with ideas or at least it serves as a vehicle for ideas. It is a Platonic matter. Prigogine has said[22]: "Matter at equilibrium is dull. The further one goes away from equilibrium, the more intelligent matter becomes."

The form of matter that I have considered in this book, living matter, is fundamentally far from equilibrium. This, too, agrees with Prigogine's notion. We could simply say "living matter," meaning that matter far from equilibrium is essentially alive. That is not a tautology, for to be alive is a physical property. Life is not an accident in the Aristotelian sense; it is not something flicked together. Life is an inherent property of a material substance and manifests itself whenever matter is far from equilibrium. In both physics and biology, then, we come back to the concept of matter due to Plato and the pre-Socratics, in which there was not yet a dualism of spirit/soul and matter.

God's Creation

The natural sciences have sought to explain the development of living things and the origin of species in terms of evolution. I have tried to show in chapters 3 and 4 that such an explanation is indeed possible. In doing so, however, we encountered the concept of self-organization, which is no longer explainable scientifically, but rather requires a new axiomatic foundation.

After the theory of evolution succeeded in accounting for the formation of life, the origin of species, and the behavior of animals and humans, there was no dearth of attempts to take one step further scientifically and to attempt to establish an evolutionary basis for the phenomena of religiosity and God; examples include the work of Alister Hardy[23] or Hoimar von Ditfurth[24]. However, these attempts are doomed to failure for fundamental reasons, because incomparable things are being compared. Reinhard Löw has given the clearest and cleverest answer to such attempts: "The empirical evidence that forms the basis of belief attests to something unique, something incommensurable with science, since it is concerned with what happens 'as a rule'. Science rests on the pillars of reproducibility and regularity. However, there is also the experience of the unique, interpersonal, aesthetic, religious experience, the experience of 'meaning.' It eludes the restricted concept of experience in the sciences and yet it is no less real than this; indeed, within the context of life, it gives the specialized form of conduct called 'science' a purpose, which by itself it does not have. It comes before all measurement and counting. To defend his thesis, Ditfurth would have to deny the authenticity of sensory impressions. He would have to interpret them in the context of a scientific worldview instead of interpreting the scientific worldview in the light of this experience. But with what authority?"

"Whom do we regard as an authority for a certain kind of experience: he who has had it or he who has no idea of it? Who is competent to judge the beauty of Beethoven's A flat major piano sonata: the unmusical expert on acoustics or the music lover, who has heard a lot, but himself is able to play nothing?. . ."

"If God created 'matter and the rules of the game,' then why shouldn't evolution occur according to his will and be in complete accord with a reasonable theory of evolution?"[25]

Religion is thus in accord with a "reasonable" theory of evolution. But what does reasonable mean? In my opinion, it can only mean that a theory of evolution cannot contain any *hidden* metaphysical elements disguising its origin.

I think that, in my discussion of self-organization, I have discovered and named the actual metaphysical element in a scientific theory of evolution. There can be no physics without metaphysical foundations, but it is extremely important that the boundary where the two meet is clearly defined in order to avoid confusions in terminology. In the theory of evolution, the concept of self-organization represents the boundary separating theory and metatheory. In the final analysis, then, the concept of matter so prevalent in the natural sciences has to be "sacrificed." And why not? It was sacrificed in particle physics long ago, although things there are so abstract that they have not yet become a part of our general awareness. Evolution could lie in God's will. It could be God's creation.

Does God still have a place in the mind of a scientist? The new concept of matter I have outlined here provides, I think, an unequivocal answer to this question: yes. The biblical story of creation is neither explained nor pushed aside by my theory of the evolutionary field. In the biblical story of creation, God reveals himself to man, not only offering an explanation of the world (as I have tried to do scientifically) but also giving the world a meaning. The question of meaning, however, is excluded a priori from scientific questions and explanations.

Matter is filled with ideas in the theory of the evolutionary field. Although it is fundamentally impossible to show whether matter has a divine element, it could serve a vehicle for the divine. This would no longer be in contradiction to the extended scientific concept of matter proposed here.

Roald Hoffmann

Autumn Entelechies[26]

1

The fever is past
but I feel fragile.
Like the Egyptian glass bottle of iridescent green,
  pasted together, but showing the cracks.
Like the Nabatean beads, peeling away sharp,
  onion-like, but corroded layers.
Like the old Coptic textile fragment, tattered
  and fading in all but its yellow and red.
I feel fragile.
My pieces are all there,
but they are held by weak ties.
My head feels the draft.
Mount me in the same museum case.
  Protect me from the wind.
    Arrange me and I will come to life again.

2

These are the days when the clouds
descend on our town. You see
them coming from our side.
The town is processed
by their passage, piecemeal
fabricated, pressed into existence.
Tree trunks made to be lost in the camouflage
of fall now jut before the fog.
That yellow house wasn't there before!
The glen's cleft protrudes.
A two dimensional curtain
focuses a plane
by obliterating the background.
Then, against your mind's
ever-conservative
wish

to freeze
that scene,
while you scan
it changes.

3
Things have such difficulty
in becoming ... The restless
blackbirds in the trees there,
what makes them so?
Too easy for the toolmaker
in me to zoom in, dissect, and
in the end (or at least
where I choose to stop)
adduce — neat molecules,
restive, stochastically
colliding to fabricate
the biochemical tinkerer's
tool kit, with it to assemble,
in sublime bondage
to the anarchy that drives,
things — as simply
laid out as microscopic
barbs on feathers, even
what is built into the chatter
of obscured birds.
But that
will not do. A purpose must be
externally organized; here
the hunter's gun, shot scattering,
reverberations — afeared,
in cawing disarray, they assail
the space newly cleared by the leaves,
are strewn to the sky ...
only, in sweet time
to wheel into the flock
that we demand them to be.

8. Aging and Death – Our Time

A dialogue between Socrates and his student Krebs on death and on life after death. The conversation took place in Socrates' death cell a few hours before his execution and is recounted by Plato[1]

KREBS: *In what concerns the soul, men are apt to be incredulous; they fear that when she has left the body her place may be nowhere, and that on the very day of death she may perish and come to an end – immediately on her release from the body, issuing forth dispersed like smoke or air and in her flight vanishing away into nothingness. If she could only be collected into herself after she has obtained release from the evils of which you were speaking, there would be good reason to hope, Socrates, that what you say is true. But surely it requires a great deal of argument and many proofs to show that when the man is dead his soul yet exists, and has any force or intelligence.*

SOCRATES: *True,* my dear Krebs, *and shall I suggest that we converse a little of the probabilities of these things?*

KREBS: *I am sure that I should greatly like to know your opinion about them.*

SOCRATES: *If you please, then, we will proceed with the inquiry. Suppose we consider the question whether the souls of men after death are or are not in the world below. There comes into my mind an ancient doctrine which affirms that they go from hence into the other world, and returning hither, are born again from the dead. Now, if it be true that the living come from the dead, then our souls must exist in the other world, for if not, how could they have been born again? And this would be conclusive, if there were any real evidence that the living are only born from the dead; but if this is not so, then other arguments will have to be adduced.*

KREBS: *Very true.*

SOCRATES: *Then let us consider the whole question, not in relation to man only, but in relation to animals generally, and to plants, and to everything of which there is generation, and the proof will be easier. Are not all things which have opposites generated out of their opposites? I mean such things as good and evil, just and unjust – and there are innumerable other opposites which are generated out of opposites.*

And I want to show that in all opposites there is of necessity a similar alternation; I mean to say, for example, that anything which becomes greater must become greater after being less.

KREBS: *True.*

SOCRATES: *And that which becomes less must have been once greater and then have become less.*

KREBS: *Yes.*

SOCRATES: *And the weaker is generated from the stronger, and the swifter from the slower.*

KREBS: *Very true.*

SOCRATES: *And the worse is from the better, and the more just is from the more unjust.*

KREBS: *Of course.*

SOCRATES: *And is this true of all opposites? and are we convinced that all of them are generated out of opposites?*

KREBS: *Yes.*

SOCRATES: *And in this universal opposition of things, are there not also two intermediate processes which are ever going on, from one to the other opposite, and back again; where there is a greater and a less there is also an intermediate process of increase and diminution, and that which grows is said to wax, and that which decays to wane?*

KREBS: *Yes.*

SOCRATES: *And there are many other processes, such as division and composition, cooling and heating, which equally involve a passage into and out of one another. And this necessarily holds of all opposites, even though not always expressed in words – they are really generated out of one another, and there is a passing or process from one to the other of them?*

KREBS: *Very true.*

SOCRATES: *Well, and is there not an opposite of life, as sleep is the opposite of waking?*

KREBS: *True.*

SOCRATES: *And what is it?*

KREBS: *Death.*

SOCRATES: *And these, if they are opposites, are generated the one from the other, and have their two intermediate processes also?*

KREBS: *Of course.*

SOCRATES: *Now, I will analyze one of the two pairs of opposites which I have mentioned to you, and also its intermediate processes, and you shall analyze the other to me. One of them I term sleep, the other waking. The state of sleep is opposed to the state of waking, and out of sleeping waking is generated, and out of waking, sleeping; and the process of generation is in the one case falling asleep, and in the other case waking up. Do you agree?*

KREBS: *I entirely agree.*

SOCRATES: *Then suppose you analyze life and death to me in the same manner. Is not death opposed to life?*

KREBS: *Yes.*

SOCRATES: *And they are generated one from the other.*

KREBS: *Yes.*

SOCRATES: *What is generated from the living?*

KREBS: *The dead.*

SOCRATES: *And what from the dead?*

KREBS: *I can only say in answer – the living.*

SOCRATES: *Then the living, whether things or persons, are generated from the dead?*

KREBS: *That is clear.*

SOCRATES: *Then the inference is that our souls exist in the world below?*

KREBS: *That is true.*

SOCRATES: *And one of the two processes or generations is visible – for surely the act of dying is visible?*

KREBS: *Surely.*

SOCRATES: *What then is to be the result? Shall we exclude the opposite process? and shall we suppose nature to walk on one leg only? Must we not rather assign to death some corresponding process of generation?*

KREBS: *Certainly.*

SOCRATES: *And what is that process?*

KREBS: *Return to life.*

SOCRATES: *And return to life, if there be such a thing, is the birth of the dead into the world of the living?*

KREBS: *Quite true.*

SOCRATES: *Then here is a new way by which we arrive at the conclusion that the living come from the dead, just as the dead come from the living; and this, if true, affords a most certain proof that the souls of the dead exist in some place out of which they come again.*

KREBS: *Yes, the conclusion seems to flow necessarily out of our previous admissions.*

SOCRATES: *And that these admissions were not unfair may be shown, I think, as follows: If generation were in a straight line only, and there were no compensation or circle in nature, no turn or return of elements into their opposites, then you know that all things would at last have the same form and pass into the same state, and there would be no more generation of them.*

KREBS: *What do you mean?*

SOCRATES: *A simple thing enough, which I will illustrate by the case of sleep. You know that if there were no alternation of sleeping and waking, the tale of* sleeping beauty* *would in the end have no meaning, because all things would be asleep too, and she would not be distinguishable from the rest. Or if there were composition only, and no division of substances,* if all things assumed a state of maximum entropy, *then the chaos of Anaxagoras would come again. And,* my dear Krebs, *in like manner, if all things which partook of life were to die, and after they were dead remained in the form of death, and did not come to life again, all would at last die, and nothing would be alive – what other result could there be? For if the living spring from any other things, and they too die, must not all things at last be swallowed up in death?*

KREBS: *There is no escape; and to me your argument seems to be absolutely true.*

SOCRATES: *Yes, it is and must be so, in my opinion,* my dear Krebs; *and we have not been deluded in making these admissions; but I am confident that there truly is such a thing as living again, and that the living spring from the dead, and that the souls of the dead are in existence.*

Classical Physics – The Exclusion of Time

Time is a unit of measurement that relates the past to the present and the present to the future. The steady passage of time, reckoned according to the motion of heavenly bodies, is a generally valid measure for the occurrence of all physical events. This, at least, is the concept of time in classical physics. After all, a major focus of classical physics was the motion of heavenly bodies and the corresponding laws of motion. It is hardly surprising, therefore, that this concept of time accords with that of "physical reality"; the concepts are made for each other.

Directly linked to this concept of time is the notion of causality, since a cause in the past produces an effect in the present and the present constellation of causes governs the future. This is a self-contained framework for viewing the world and should not be shaken needlessly. It provides a sense of security, the feeling of being situated within a definite course of events and of belonging there. Indeed, causality is a categorical prerequisite for the conduct of science. The concept of time in classical physics thus represents a kind of ruler, laid against a sequence of (moving) events. It is possible, in principle, to lay this ruler against any sequence or even to turn it around and place it in the opposite direction. Physical time can run backward – that is, it can be counted backward – and processes are repeatable. We therefore regard the

* In Plato's dialogue, Socrates actually mentions Endymion, a (male) sleeping beauty of Greek mythology.

structure of time as symmetrical. But we should not forget that classical mechanics reduces the concept of time – necessarily – to what is essential for the description of mechanical motions. The description of living things, however, is entirely different. Even Lichtenberg pondered this problem: "If a person, after reaching one hundred years of age, could be turned around like an hourglass and become younger, while still exposed to the usual risk of dying, what would the world be like?"

Without "time" we could not imagine the occurrence of any process. In fact, because everything in the world consists of processes, we could not imagine anything at all. For this reason Kant defined time (and space) as a priori forms of intuition, as pure intuitions, the prerequisites of knowledge.

Now, classical physics examines only a relatively narrow slice of nature. It is not "the science," the science of all nature. Since its primary concern is mechanical motion, it views the world as stable and reversible. The world is a windup clock that, in principle, can also run backward. Of course, the explanations made possible by this concept of time are astonishing. They range from the laws of planetary motion to the theory of relativity. But we now know that classical physics, even including the theory of relativity, describes only a part of our physical world – namely, objects whose masses and energies are of the same order of magnitude as our own. The most important universal constants restrict the validity of the classical laws of motion. One example is Planck's constant (6.6×10^{-27} erg sec), the smallest possible "quantum of action"; there is no smaller impulse. Another is the speed of light (3×10^{10} cm/sec); there is no higher speed. This means that the laws of motion are not applicable beyond this speed[2].

Even at the peak of influence of Newtonian mechanics on 18th-century thought, far sighted philosophers discerned the self-restricting nature of classical physics. For instance, Diderot, in a fictitious dialogue with d'Alembert, wrote: "Do you see this egg here? With it one can turn all schools of theology and all places of worship on earth upside down. What is this egg before the seed enters it: a senseless mass. . .But how does this mass acquire a different structure, the ability to feel, life? Through heat. How is heat generated? Through motion. What are the successive effects of motion? Do not give me an answer, but take a seat instead. We will examine them exactly, from one moment to the next. First of all, there is an oscillating point, then a tissue that spreads out and becomes colored; afterwards, flesh is formed. A beak, rudimentary wings, eyes, claws; a yellow mass separates and forms the entrails. Now it is an animal. . .It hatches, it walks, it flies, it responds, it runs away, it comes closer, it cries, it suffers, it loves, it is loved, it enjoys itself. It has all of its characteristics. It performs all its activities. Do you mean to assert, like Descartes, that it is but a machine to be imitated? Then children will poke fun at you and philosophers will reply: If it is a machine, then you, too, are one. Admit, though, that the only difference between an animal and yourself is the organic structure; you display understanding and reason and are therefore on the right path. In contrast to what you think, however, one is compelled to conclude from all this that an inactive matter with certain tendencies, once it is penetrated by another inactive matter, by heat and motion, can give rise to everything: the faculty of sensation, life, memory, consciousness, passions, thought...Listen to your own words and you will regret them: You will realize that you are sacrificing common sense, because you refuse to accept a simple principle that explains everything, namely, the faculty of sensation as a general property of matter or as a product of organic structure. Therefore, you plunge into an abyss of mysteries, contradictions, and absurdity."[3]

Through this fictitious conversation, Diderot wanted to show that not all phenomena in nature are subject to treatment by a Newtonian approach. He exemplified this by describing the formation of a chick from an egg. Diderot refers to the faculty of sensation as a general property of matter. He, too, already recognized, as I discussed in chapter 7, that heaviness or inertia does not constitute an exhaustive description of matter whenever processes are involved.

Time and Entropy – "Processual" Time

The law of entropy, the first natural law attempting to describe processes, placed a stumbling block in the perfectly ordered world of classical physics. According to this law, entropy necessarily increases until the world ends in thermal death, that is, until all differences in energy have been leveled, all balls have rolled into their holes, all things have decayed and become uniform, homogeneous, featureless, and monotonous. But what is the justification of the law of entropy? After all, there are forms and structures, there is life, there are improbable states, there are states far from equilibrium.

At the "moment" of thermal death, nothing further happens. Nothing can be measured. The concept of time becomes meaningless, just as meaningless as the notion of a "moment" before the Big Bang. To describe the course of real events in this world, then, one might define an entropic time: entropy constantly increases according to the second law of thermodynamics and this increase is a measure of time. This time – our time – is irreversible.

There has been no lack of attempts by renowned physicists to trace the second law, the law of entropy, to classical mechanics. The temperature of not only a gas but also a liquid or solid is governed by the motion of its component particles. As matter becomes hotter, the random motion of the individual particles increases. The water molecules in a snowflake, for example, move about so violently above a certain temperature (0 degrees Celsius) that the snowflake melts. On further warming, the turmoil of molecules in the liquid steadily increases, until they finally start to escape from the liquid and vaporize. Under normal conditions this occurs at 100 degrees

Celsius. Heat, then, is nothing more than the motion of particles. The law of entropy states that heat eventually becomes evenly distributed; after many collisions, all particles move at roughly the same speed. This is the state of thermal death.

In this state of uniform energy distribution, which corresponds to complete and final equilibrium, the motion of the particles should be describable, in principle, by Newtonian mechanics; accordingly, the law of entropy ought to be interpretable in terms of molecular kinetics (H theorem). The first effort in this direction was made by Ludwig Boltzmann[4].

His attempt to trace the law of entropy and hence the concept of time to the classical laws of motion should be regarded as a failure today. Karl Popper has commented: "I find Boltzmann's idea breath-taking in its cleverness and beauty. But I also find it completely untenable, at least for a realist. It makes an illusion of change that occurs in only one direction. Yet this makes the catastrophe of Hiroshima an illusion as well. It makes an illusion of our world and *thus of all our attempts to learn more about our world.*"[5]

In a certain sense one could say that, other than as a scalar unit of measurement, time has been excluded from science by Newtonian physics. Furthermore, since most of mankind is compelled to live with science and its technological consequences, time has been excluded from the world as well. A Newtonian world is ageless – a world without time. Here, once again, we should recall the famous question posed by the Swedish Royal Academy in 1890 and given an answer by Poincaré: How stable is our planetary system? According to Newton, it is indefinitely stable. But post-classical modern science deals with complex processes, with the formation of the universe, with the origin of life, with the temporal nature of evolution (a process of historical development!), with the weather, with the functional processes of living organisms, with medical problems – and, in such a science, the concept of time has to be reconsidered. Evolution takes place in time. The evolutionary field, as I have proposed it, is four-dimensional with time as the fourth dimension.

Let us again consider time starting from the Big Bang. The expansion of the universe originates from the Big Bang singularity, a state of infinite density in which at first nothing happens, perhaps only for an infinitesimal fraction of a second or, to be more precise, 10^{-43} seconds, the so-called Planck time. It is meaningless to ask what happened prior to the Big Bang singularity, the beginning of space and time. It represents a "cosmic horizon" where time rises and where, possibly after a contraction, time and the universe will someday set[2].

To try to conceive of time beyond this horizon is pointless. What came before, where time comes from, what its "horizon" is – these questions cannot be asked. What "causes" the law of entropy, according to which time is irreversible, is not a physical but a metaphysical question, just as it is a metaphysical question to ask why matter is heavy or what a gravitational field "really is". Bernd O. Küppers[6] has pointed out that it is easy to become trapped in a vicious circle here. On the one hand, time is a phenomenon expressible in the form of an empirical statement, the second law of thermodynamics. On the other hand, it is evident – and has been so since Kant – that the structure of time is a prerequisite of empirical knowledge, making possible the very empirical knowledge formulated in the second law. How can one escape from this vicious circle of reasoning?

The answer is as follows: Only through consistency. It is utterly futile to attempt to derive the structure of time from underlying physical laws. Rather, we must establish the consistency of the following assumption: if the second law is a generally valid natural law, then time *can* have the structure described by us. Carl Friedrich von Weizsäcker, who first explored these ideas, carried out the consistency proof by using a more explicit version of the concept of probability[7].

The irreversibility of a natural process is found to be consistent with the time symmetry of the underlying mechanical laws if, in the statistical interpretation of the law of entropy, probabilities are calculated only for *real* transitions, that is, for transitions into the future. The first consequence is that entropy increases in the direction of the future. However, because every bygone moment was once present, the consequence is an increase in entropy for everything then lying in the future, including times that are now part of the past.

In this consistency proof, Carl Friedrich von Weizsäcker employs only the assumption that the concept of probability can be applied meaningfully solely to future (that is, possible) events, because it is meaningless to inquire into the probability of past (factual) events. The time-asymmetric application of the probability concept, which is grounded in the reality of documents, is the point at which the structure of time enters into the statistical justification of the second law.

Let us summarize the conclusions reached so far. The structure of time, as it is manifested in the difference between past and future, is not deducible from the fundamental laws of physics. Rather, it must be present a priori as an objective property of natural events.

If we argue in this way – that is, if we regard the structure of time as an objective fact that is responsible for the irreversibility of natural processes – then, as Michael Drieschner[8] has pointed out, the reversibility in the theories of classical physics becomes the phenomenon in need of explanation. Irreversibility is exactly what one would expect.

Indeed, it can be shown that the anisotropy of time leads to irreversibility in the fundamental theories of physics. This anisotropy underlies the principle of causality, since a causal statement – say, "Lightning is followed by thunder" – already assumes the structuring of time into modes such as "after" and "before." The causality of natural events would be violated, however, if it were *not possible* to take a process that obeys natural laws and, by applying the following three operations (the so-called CPT theorem), arrive at another natural process that equally obeys the same natural laws:

C: reversal of all *c*harges;

P: reflection through a *p*oint in space;

T: reversal of all motions in *t*ime.

This means that the elementary natural laws are consistent with the time structure of reality and the causality principle based on it only at the expense of a symmetry that interchanges past and future[9].

The compelling conclusion is that science is incapable of providing a deeper interpretation of the structure of time. In acknowledging this, however, we have already taken the path of philosophy.

We said that the anisotropy of time is a prerequisite of the causality principle and thus an essential prerequisite for the acquisition of empirical knowledge. This statement brings us closer to the Kantian thesis that the forms of our empirical knowledge are dictated by time and space, which exist a priori and indeed predicate the very possibility of such knowledge.

This could be regarded as an approach to a purely subjective philosophy of time, in opposition to the scientific attempt to explain the structure of time objectively. This apparent contradiction vanishes, however, when the Kantian thesis is viewed in the light of evolution. Our "worldview apparatus" – namely, the human brain and the specific faculties of human consciousness – represents the culmination of a long phylogenetic development. The brain is assigned the task of gathering information on the outside world, storing it, and transforming it into appropriate strategies for ensuring our survival. Our worldview apparatus can accomplish this task only if the picture it paints truly reflects the structure of the external world. With respect to the

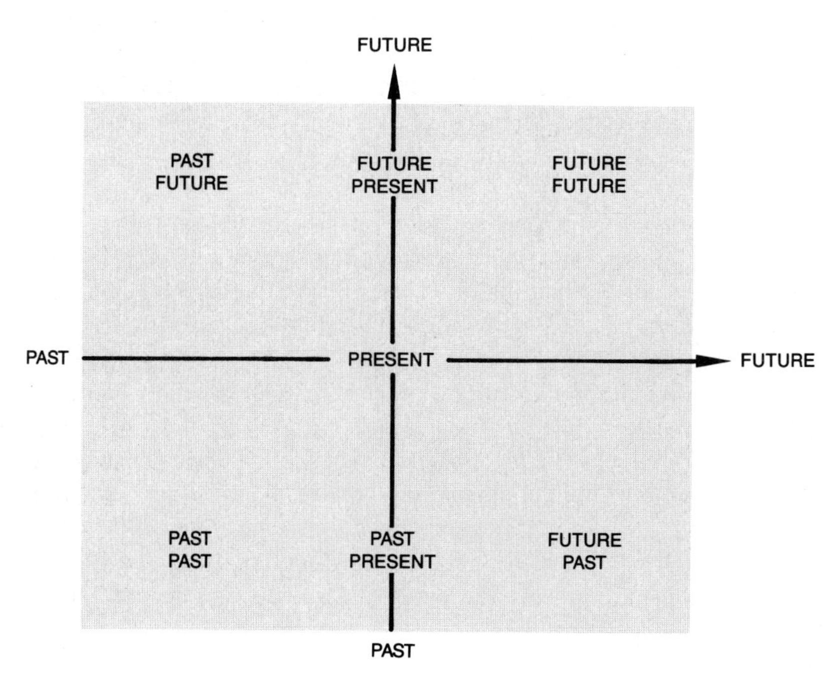

Fig. 8.1. Superposition of the modes of time according to Bernd O. Küppers.

evolutionary adaptation of an organism to its physical environment, our consciousness of time and thus our causal way of thinking is only in accord with a reality in which time is objectively asymmetric.

However, this conclusion also requires a consistency proof, since our scientific theories, particularly the theory of evolution, are themselves products of our worldview apparatus, our brain. Indeed, we seek to describe the structure and function of the brain with the help of this very theory. In the light of evolution, then, the Kantian a priori of the individual emerges as an a posteriori of evolutionary history. Our intuitive perception of time, in turn, is accessible to an objective analysis within the context of the biological sciences[6,10].

Unquestionably we possess a very complex consciousness of time, as expressed in the one-dimensional arrangement of past, present, and future. But does the assumption that time has a one-dimensional physical structure really enable us to understand the total extent and complexity of our consciousness of time?

A solution to this problem emerges when we make time itself the object of experience, the living experience of human beings. If the thesis developed so far is correct, namely, if empirical knowledge is only possible in our temporally structured reality, then the experience of structured time – that is, the experience of past, present, and future as a process in time – must itself have a time structure. Accordingly, it is useful to superimpose the modes of time and to speak, for example, of the past of the present, the present of the present, and the future of the present[11]. This holds for the other modes of time as well (past and future), resulting in a total of nine first-order superpositions (Fig. 8.1).

Now, the being that manifests itself in the first-order superpositions is, in turn, a being in time and must therefore appear in the trinity of time's modes. This results in 27 second-order superpositions. This process of superposition can be reiterated at will, resulting in a homogeneous, multidimensional time structure with no upper limit to the number of dimensions. The one-dimensional projection of this multidimensional structure gives physical time in the trinity of its modes: past, present, future. Only this physically objective part of the time structure is accessible to our senses and hence measurable.

The superposition of the modes of time lends a markedly hierarchical fine structure to the one-dimensional structure of time; this fine structure manifests itself only in the organization of our consciousness. A systematic pursuit of this approach leads to the thesis that being and time are not independent of each other; being *is* time. Being is not merely the substrate acted on by time; it constitutes itself through time (Martin Heidegger).

And thus we arrive at a definition of time that King Solomon gave three thousand years ago[12]:

To every thing there is a season,
and a time to every purpose under the heaven:
A time to be born, and a time to die;
A time to plant, and a time to pluck up that which is planted;
A time to kill, and a time to heal;
A time to break down, and a time to build up;
A time to weep, and a time to laugh;
A time to mourn, and a time to dance;
A time to cast away stones, and a time to gather stones together;
A time to embrace, and a time to refrain from embracing;
A time to seek, and a time to lose;
A time to keep, and a time to cast away;
A time to rend, and a time to sew;
A time to keep silence, and a time to speak;
A time to love, and a time to hate;
A time for war, and a time for peace.

Let us recapitulate this difficult chapter on processual time. Time is irreversible and is a measure of the course of evolution in the evolutionary field. From conception to death we are embedded in this course of evolution. The structure of time is directional and irreversible. In the Kantian sense, however, it is closely related to the reversibility of the fundamental theories of classical physics, which themselves are the exceptions requiring explanation. The formation of the new and the decay of the old occurs in irreversible time.

Aging and Death – A Biochemical Problem?

Each form of higher life has a characteristic life span. The biblical life span of man is seventy to eighty years, as stated in the 90th Psalm: "The days of our years are threescore and ten, Or even by reason of strength fourscore years; Yet is their pride but travail and vanity; For it is speedily gone, and we fly away." Modern medicine and hygiene might prolong this age by perhaps ten years. But even with further improvements in medicine and geriatrics the average life span will not extend much beyond one-hundred years. The human survival curve (Fig. 8.2) indicates that, although more and more people are attaining a maximum age of between ninety and one-hundred years, this maximum age is not increasing. This maximum life span must be preprogrammed in some way, since even closely related species in the animal kingdom often have completely different average life expectancies.

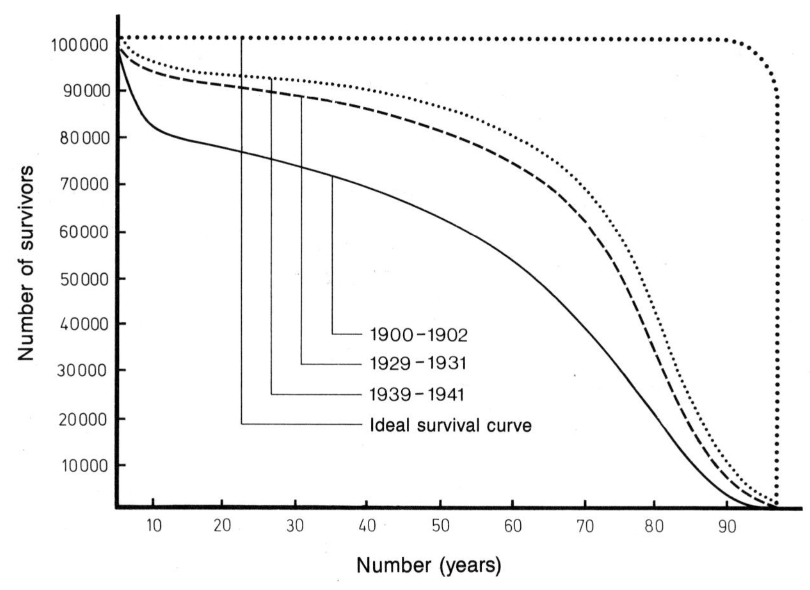

Fig. 8.2. Number of survivors per 100000 human births in the USA since the beginnings of modern medicine[13].

Presumably, then, some kind of genetic program governs aging or, better, specifies a strategy to delay aging, to avoid chaos. This strategy has been developed to greater or lesser extents in the various species. After all, the difference cannot lie in the basic biochemical mechanisms, since they are the same in all species and organisms: DNA carries the genetic information written in the universal genetic code; this code is translated into proteins, which, in turn, catalyze all anabolic and catabolic reactions in the organism through virtually the same enzyme mechanisms.

The genetic basis of aging is presumably very complex. Yet it is certain that the genome itself – that is, the DNA carrying the genetic information – does not grow old. Otherwise, older parents would give birth to older, perhaps even senile, children. A sixty-year-old father and a forty-year-old mother would produce a child that, at birth, has the "biochemical age" of a fifty-year-old. This is nonsense, of course. Aging is genetically preprogrammed, to be sure, but it is an "epigenetic" process[14], one based on additional information introduced during differentiation.

Does biochemical aging represent an aging of the DNA? An "aging" of nucleic acids, that is, a chemical change, corresponds to a mutation, as we saw in chapter 2. Each change in a single nucleic acid building block is, in principle, detectable as a mutation, even if it represents only *one* change in the human genome of 10^{10} building blocks. However, what we have just said – namely, that the babies of older parents show no signs of aging – unequivocally shows that the nucleic acids of the germ cells (the female ovum and the male sperm cell) do not age biochemically. If

Species	Average life expectancy (years)	Species	Average life expectancy (years)
Fly	0.077	Rooster	20
Mouse	3–3.5	Tiger	20
Rat	3–3.5	Lion	20–25
Rabbit	5–7	Cattle	20–25
Guinea pig	8	Ape	20–30
Cat	9–10	Horse	20–30
Fox	10	Pig	20–30
Squirrel	10–12	Camel	40–50
Dog	10–12	Crocodile	50
Ant	10–15	Carp	50–60
Frog	10–15	Falcon	60–70
Sheep	10–15	Raven	60–70
Goat	12–15	Man	70–74
Wolf	12–15	Galapogos tortoise	100–150
Herring	16	Elephant	150–200

aging is associated with the DNA, then the nucleic acids in differentiated somatic cells would be expected to mutate or decay faster than the nucleic acids in germ cells. There is no evidence and indeed no good reason for this. Certainly, epigenetic changes in the nucleic acids do occur – cancer may be one example and could perhaps be defined as a symptom of old age or of the degeneration of cells – but these changes are unquestionably pathological in nature and have nothing to do with aging per se. Aging, then, cannot be ascribed to "diseased" nucleic acids; instead, it must be regarded as an epigenetic phenomenon, albeit one that is programmed by the nucleic acids, since the average life expectancy is genetically determined.

Let us take a look at the possible epigenetic, biochemical events that might be responsible for aging. Defined biochemically, an organism is the sum or the network of its proteins and their interactions. Accordingly, aging ought to reflect the aging of proteins. This is substantiated by experimental findings. Simply examine the skin of Dürer's mother in the portrait shown in Figure 8.6. Skin consists of a network of proteins and these clearly change with age. This change has been studied in detail for collagen, the main component of connective tissue[15].

Are these changes preprogrammed, then, or are they due to gradual and accidental degeneration? We do not know. Outwardly, the ensemble of proteins making up the organism appears static to us. A person's appearance changes only slowly. In reality, though, every organism represents a dynamical state in which everything is flowing, as portrayed poetically by Conrad Ferdinand Meyer in his poem "The Roman Foun-

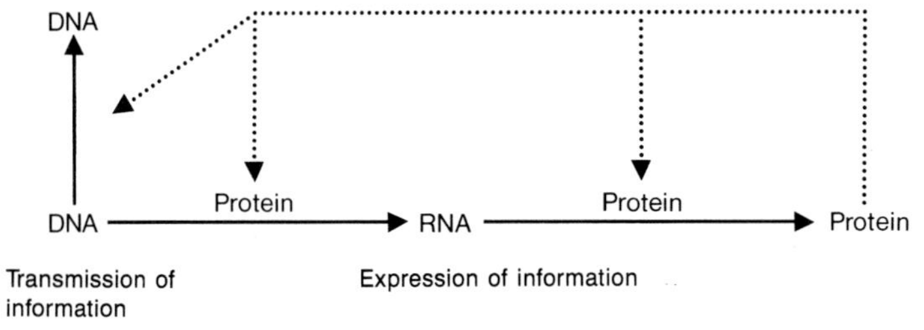

Fig. 8.3. Error feedback in protein biosynthesis. The genetic information is first transcribed into ribonucleic acids, which, in turn, are translated into the "correct" protein sequence. Because the resulting proteins are involved in their own synthesis, errors are potentially subject to rapid amplification.

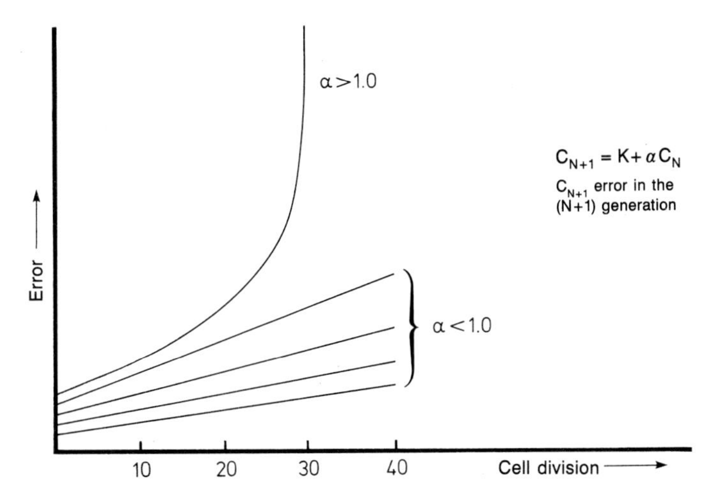

Fig. 8.4. Error catastrophe as proposed by Leslie E. Orgel.

tain" (see p. 131). In these dynamical processes of anabolism and catabolism, proteins are newly synthesized, as we have seen in chapters 1 and 2. However, protein synthesis is afflicted by a ingrained source of error; the whole system of the organism is subject to a high degree of feedback (Fig. 8.3).

The information for the synthesis of proteins is taken from the genetic storehouse. These instructions direct the synthesis of proteins, which, in turn, are involved in the synthesis of other proteins. If an error, even a small one, is made somewhere in this system, feedback amplification can occur rapidly, leading ultimately to an error catastrophe. This is the essence of the error catastrophe theory of Leslie E. Orgel[16].

The number of errors in each successive generation is expressed by the simple equation shown in Figure 8.4, assuming, for the sake of simplicity, that a linear differential equation can be set up for this relation. According to this equation, the number of errors in a given generation is equal to a constant K plus the number of errors in the preceding generation multiplied by a factor α. If α is greater than 1, this simple equation leads to an error catastrophe. Using various methods, we have measured the error rate of protein biosynthesis (see chapter 2). In addition, we have discovered the error-avoidance strategy employed by the machinery of protein biosynthesis, as discussed in chapter 2[17,18].

This ingenious strategy offers the further advantage that it involves a dynamical dissipative structure. By regulating the supply of energy, therefore, an organism can largely suppress the error rate and, in the ideal case, reduce it to only one error in one million protein building blocks.

But is this "ideal case" really ideal? An organism is capable, in principle, of virtually flawless protein synthesis, if necessary at the expense of large amounts of energy. The question is whether the effort is worth it. Perhaps a flawless protein is not ideal, after all. Perhaps the gradual degeneration of proteins serves a useful purpose, since it is not all that desirable for an individual organism to have an endless life span (see below).

How stable should our genetic system be? If the feedback catastrophe is indeed intentional, then the life span should be governed by the average inherent rate of error in protein biosynthesis. In this case, though, the scatter of life expectancies would be substantially larger than that predicted by the simple Orgel theory.

At numerous points in this book, we have been introduced to systems subject to feedback. It may very well be that aging corresponds to a more complex feedback system than the equation given in Figure 8.4 would indicate; possibly the Verhulst equation provides a better description of this process (see chapter 5). This would mean that, beyond a certain error rate and for a certain rate of synthesis, protein biosynthesis becomes a chaotic system. At a threshold value, the biosynthetic machinery of the cell suddenly and catastrophically breaks down. Perhaps this is what is meant by "biochemical death."[20]

But we still know much too little. We simply do not have sufficient data on the individual parameters and rates of protein biosynthesis, nor can we penetrate deeply enough into the system character of this complex process. In any event, the final collapse of an organism must be a sudden event, since it is difficult, for example, to detect biochemical changes in the proteins of old mice and rats[19].

The Verhulst character of such an event might explain the lack of success so far in the effort to detect the *gradual* decay of proteins. Owing to the chaos-avoidance strategy of this dynamical structure, the catastrophe occurs suddenly. But we still lack sufficient data to perform calculations on such systems. Some of what we do know is summarized below.

The situation is probably much more complicated, however. In higher organisms, the network character of the overall system has to be taken into account. This is illustrated in Figure 8.5.

Not only can protein biosynthesis end in collapse as a result of feedback catastrophe. A large number of specific processes are dependent, in turn, on the proper func-

Possible basis for performing calculations on the network of life

1. Experimentally determined error rate for the incorporation of amino acids

Yeast	Ile/Val	$1:3.8\times10^4$
E. coli	Phe/Tyr	$1:3.7\times10^5$
	Phe/Leu	$1:9.2\times10^5$
	Phe/Met	$1:7.6\times10^5$

2. Lifetime of important proteins (measurable)
3. Effect of errors on function (determinable from mutants)

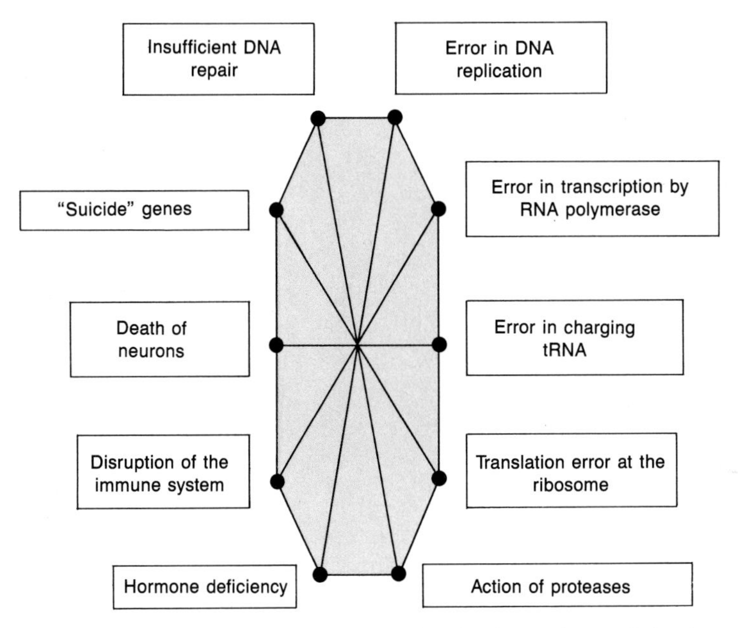

Fig. 8.5. Network of molecular events that lead to aging and death.

tioning of proteins. One example is the synthesis of hormones, proteins that have positive or negative effects on the entire organism and which, by influencing the overall physiology, enhance or weaken other molecular effects. Or the immune system, which normally protects the body from infections and degenerate cells. If damaged – when a person has AIDS, for example – it no longer prevents the body from rapidly decaying. If not functioning properly, it leads to autoimmune diseases in which the body attacks itself.

All the functions shown in Figure 8.5 interact and regulate one other; in principle, they can form a system, which, at an abrupt threshold, collapses[20,14]. Accordingly, microscopic disorder could build up to macroscopic catastrophe – to death.

In this book, we have been introduced to numerous dynamical networks subject to feedback, similar to those in Figures 8.3 and 8.5. It is these dissipative structures that make possible the creation of order and are indeed responsible for the unimaginably precise functioning of living systems. Why? Because they operate on the edge of chaos. For this very reason they are inherently subject to sudden collapse; they carry within themselves the seeds of their own death. If there is such a thing as a "universal formula for life," it would be the basic equation of Figure 6.2 on p. 138. This formula summarizes – admittedly in a very general way – most of the processes of life discussed in this book, all of which involve self-organizing matter.

Aging – Fate or Disease?

In considering the phenomenon of aging, even biochemists shouldn't overlook the classical work of Seneca on the transience of life: "De brevitate vitae."[21] Here are just a few quotations: "So it is: the life we receive is not short but we make it so; we are not ill provided but use what we have wastefully. Kingly riches are dissipated in an instant if they fall into the hands of a bad master, but even moderate wealth increases with use in the hands of a careful steward; just so does our life provide ample scope if it is well managed."

"If is because you live as if you would live forever; the thought of human frailty never enters your head, you never notice how much of your time is already spent. You squander it as though your store were full to overflowing, when in fact the very day of which you make a present to someone or something may be your last. Like the mortal you are, your are apprehensive of everything; but your desires are unlimited as if you were immortal."

"It takes a great man, believe me, and one who rises high above human frailty, to allow none of his time to be frittered away; such a man's life is very long because he devotes every available minute of it to himself."

"See how eager they are for a long life. Decrepit old men beg and pray for the addition of a few more years; they pretend they are younger than they are; they flatter themselves with a lie and are pleased with their deception as if they were deluding Fate at the same time."

"The philosopher's life is therefore spacious. He is not hemmed in and constricted like others. He alone is exempt from the limitations of humanity; all ages are at his service as at a god's. Has time gone by? He holds it fast in recollection. Is time now present? He utilizes it. Is it still to come? He anticipates it. The amalgamation of all time into one makes his life long. But for those who are oblivious to the past, negligent of the present, fearful of the future, life is very short and very troubled."

In light of these classical quotations and what was said on p. 194f, the reader will have little doubt about my answer to the question whether aging is fate or sickness[22]: Aging is fate, it is biological necessity, and it is also a fundamental structural feature of an evolving world made up of dissipative structures. Everything grows old and dies: fixed stars, minerals, cells, living things, and systems of living things. Why should human aging be an exception, an aggravating illness that will be cured one day through the advances of modern medicine – what an utterly senseless notion! Naturally, it is hard for a human being, as a consciously living individual, to imagine the demise of his spirit, his consciousness – it is easier, in fact, to imagine the decay and demise of his body, of matter, of inert matter.

As I tried to show in chapter 7, our concept of matter must be revised in the light of evolution: matter is creative. On aging, inert matter deteriorates and, on death and decay, it merges with the overall stream of matter. What happens to creative matter? Presumably, after traversing zones of chaos, it is ordered once again in the evolutionary field. But this represents a step beyond the ideas at hand and enters the realm of metaphysics.

Dürer's Mother – Thoughts on the Dignity of Old Age

To conclude this chapter on "Aging and Dying," I would like to make some general observations. Let us examine the well-known drawing of Albrecht Dürer's mother (Fig. 8.6). Albrecht Dürer sketched this portrait in 1514 when he was 43 years old and his mother 63. At first glance, this wondrously sympathetic and at the same time anatomically exact drawing reveals how aged and worn the woman was at 63 years of age. Life was harder then and 63 years represented a venerable old age.

Let us, first of all, examine the drawing from the diagnostic viewpoint of medicine and point out the signs of old age. We note that the haggard skin is covered with folds and wrinkles. With increasing age the elastic properties of the subcutis change as a result of characteristic changes in the compositions of the corresponding proteins. Viewed in this way, aging is the aging of proteins. We notice further that the old woman has lost her teeth and has sunken cheeks. Dentures were unknown at that time and Dürer's mother probably suffered from periodontal disease and tooth decay. Since her resistance to bacteria had weakened in old age, the bacteria in her mouth had eaten away at her teeth. Finally, we can diagnose a third disease of old age with some certainty. The right eye is set unnaturally; it looks askance. Very likely, the old woman had glaucoma, characterized by increased interocular pressure and

Fig. 8.6. Dürer's mother. Drawing by Albrecht Dürer, 1514.

leading eventually to blindness. This is a disease of old age due to inadequate control of interocular pressure; that is, it is a regulatory disorder. Accordingly, aging is, among other things, the following:

1. Aging of proteins;
2. Weakening of resistance toward infections;
3. Regulatory disorders.

But the medical viewpoint is not the only way to examine this face. Even an experienced doctor and diagnostician sees more than these direct symptoms. The strongest and most pronounced impression given by the drawing is the *dignity* of old age. To grow old, to age with dignity in the face of death can be one of the greatest goals of a properly lived human life. In his "Maximen und Reflexionen" Goethe said: "To grow old means to start a new business; all relationships change and either one must stop acting altogether or take on the new role with determination and conscious awareness" (compare Martin Gregor-Dellin[23]).

What wonderful poems we owe to the almost seventy-year-old Goethe – the poem "Westöstlichen Divan," the poem "Um Mitternacht" from "Klein – kleinen Knaben," the "Dornburger Gedichte", the "Marienbader Elegie," and, shortly before his death, the second part of "Faust," a work par excellence of old age and yet so fresh, so full of ideas, so unconventional – a new beginning and a culmination in one. Perhaps Faust's death was, in fact, Goethe's *idealized* description of his own death, reflecting, at the same time, his skeptical, questioning attitude toward death.

Aging brings with it an unimagined freedom. It is no longer necessary to have ambitions, to sell oneself. Life proceeds, so to speak, voluntarily. Filled with all the experience of past decades, one lives one's life at once exemplarily and unconditionally. Confucius said: "Only when I turned seventy could I follow the impulses of my heart."

Whoever, like I, has been so fortunate as to have heard the nearly blind Ernst Bloch present his oratorical visions, whoever has experienced Karl Jaspers in his seminars, whoever was privileged to speak with Werner Heisenberg at the end of his life, whoever was able to exchange thoughts with the virtually transfigured and yet so youthful, close-to-life Karl Rahner, that person knows what creative fires are unleashed by the wisdom of old age, the experience of a full life, and the closeness to death. Here, it is literally true that the spirit creates the body. Michelangelo, Leonardo da Vinci, Tizian, Rembrandt, Handel, Bach, Goethe, Goya, Menzel, Verdi, Thomas Mann – all are examples of how the creative power need by no means needs flicker in old age and how, especially in the face of death, revolutionary new beginnings are attempted: "With a premonition of such elevated happiness, I now enjoy the supreme moment" said Faust and fell back dying – happiness and delusion at the same time. Fontane, himself old, drew a wonderful picture of old age in his description of Stechlin late in life. The pastor Lorenzen, standing at the grave of Stechlin, says: "When one looked at him, he seemed to be an old man, also in the way he accepted time and life; but for those who knew his true being, he was not aged, and of course not young either. Rather, he possessed that which is beyond all temporal things, which is always true and will be so always: a heart." And further: "He was quite free indeed." Fontane said of his father: "He was at the very end what he truly was."

Order, the generation of forms, the power of creation are the results of an inherent avoidance of chaos, in the universe as well as in the life of the individual. The world is dynamical; it moves, it develops. As a result, parts of the whole, despite all strategies of chaos avoidance, fall time and again into chaos. The individual life, too, ultimately ends in chaos – the individual is then dead. Yet the overall world always remains non-chaotic and creative. In the true sense of the word, chaos is a side effect. At the edge of the world things die. There, the law of entropy holds. But the creative world, the world of the creator, the evolving world of the Big Bang, the spiral nebulae, the planetary systems, the primordial soup, the spiritual – this world is eternal. Death is always individual only; it is the return to the eternal whole. From the chaos at the edge of the world order emerges. The world nurtures itself from its dead, who are absorbed back into it.

William Butler Yeats

Sailing to Byzantium[24]

I

That is no country for old men. The young
In one another's arms, birds in the trees,
– Those dying generations – at their song,
The salmon-falls, the mackerel-crowded seas,
Fish, flesh, or fowl, commend all summer long
Whatever is begotten, born, and dies.
Caught in that sensual music all neglect
Monuments of unageing intellect.

II

An aged man is but a paltry thing,
A tattered coat upon a stick, unless
Soul clap its hands and sing, and louder sing
For every tatter in its mortal dress,
Nor is there singing school but studying
Monuments of its own magnificence;
And therefore I have sailed the seas and come
To the holy city of Byzantium.

III

O sages standing in God's holy fire
As in the gold mosaic of a wall,
Come from the holy fire, perne in a gyre,
And be the singing masters of my soul.
Consume my heart away; sick with desire
And fastened to a dying animal
It knows not what it is; and gather me
Into the artifice of eternity

IV

Once out of nature I shall never take
My bodily form from any natural thing,
But such a form as Grecian goldsmiths make
Of hammered gold and gold enamelling
To keep a drowsy Emperor awake;
Or set upon a golden bough to sing
To lords and ladies of Byzantium
Of what is past, or passing, or to come.

9. Fundamental Complexity – Intrinsic Limitations

A dialogue between Georg Christoph Lichtenberg and Prince Hamlet* on the intrinsic limitations and the insular nature of our knowledge and existence

HAMLET: *There are more things in heaven and earth,* Professor, *than are dreamt of in your philosophy.*

LICHTENBERG: *You are somewhat of a saucy philosopher, Prince Hamlet, to assert that there are many things in heaven and on earth that are not at all found in our compendium of knowledge. If your intention is to cast aspersions on our compendia of physics, then the unhesitating reply is: all right, but our compendia also contain many things present neither in heaven nor on earth.*

It is not necessary to resort at once to miracles, to irrationalities, to the ghost of your lord father. In actuality, the world is fundamentally complex. The word "actuality" itself expresses this property. The world acts, it is never still, everything is interrelated. *Thus, when something acts on us, the effect is dependent not only on the thing acting, but also on what is being acted upon. Both are involved – as in a collision – and both are affected; for it is impossible for one being to suffer the effects of another without the overall effect appearing mixed. I should think that, in a direct sense, a simple tabula rosa would be impossible, because every effect modifies the agent and what it thereby loses is transmitted to the other, and vice versa.*

HAMLET: (aside) *How absolute the knave is!* (then aloud) *We must speak by the card, or equivocation will undo us. By the Lord, this three years I have taken note of it; the age is grown so picked that the toe of the peasant comes so near the heel of the courtier, he galls his kibe. What I mean is that the time is out of joint: O cursed spite, that ever I was born to set it right!* All differences, all order has been lost. All things become so hopelessly tangled, chaotic, and complex when one examines them closely.

* Prince Hamlet of Denmark, the principal character in William Shakespeare's tragedy, written in 1601.

LICHTENBERG (aside): *If this is philosophy, then at the very least it is an inane one.* (then aloud) Your Royal Highness, to set time right nowadays requires government and scientists. We physicists have known, at least since Heisenberg's uncertainty principle, what proximity and perspective mean for the analysis and solution of a problem. *Proximity is of no help to us, for the thing we are able to approach is not what we want to approach. When I gaze at the setting sun and take a step toward it, I come closer to it, even if only by a tiny amount. For the organs of the soul the situation is entirely different. Indeed, it could be that by approaching too closely – say, with a microscope – one only distances oneself further from what one is able to approach. For example, I perceive a strange mass on a mountain in the distance. I approach more closely and find that it is a castle. Still closer, I discover windows, etc. That would suffice; if I were unfamiliar with the purpose of the whole and examined it still further, I would end up analyzing the stones, which would only divert me.* In your attempt to find a all-embracing perspective, Prince Hamlet, you make life difficult for yourself. There are no all-embracing solutions. It is not necessary for a scientist to know and explore everything; but one does require precise knowledge of the perspective, the region, indeed the island of investigation. *Anybody who seeks to survey the history of philosophy and natural science will find that the greatest discoveries were made by people who considered merely probable what others held for certain, that is, by adherents of the New Academy, which maintains a balance between the rigid steadfastness of the stoic and the uncertainty and disinterest of the skeptic. I would recommend such a philosophy to you.*

HAMLET (passionately): *Words, words, words.* (Then, in despair, covering his face with his hands) *My words fly up, my thoughts remain below: Words without thoughts never to heaven go.* (After a pause, regaining his calm) *What a piece of work is man! how noble in reason! how infinite in faculty! in form and moving how express and admirable! in action how like an angel! in apprehension how like a god! the beauty of the world! the paragon of the animals! And yet, to me, what is this quintessence of dust? man delights not me!*

Oh God, I would be bounded in a nut-shell and count myself a king of infinite space, were it not that I have bad dreams.

LICHTENBERG: *Which dreams indeed are ambition; for the very substance of the ambitious is merely the shadow of a dream.*

HAMLET: *A dream itself is but a shadow.*

LICHTENBERG: (quoting Shakespeare) *Strange to relate, but wonderfully true that even shadows have their shadows too.* (Then aside) *I, too, know no one I can confide in, not even a dog whom I can call "you." In these circumstances, however, I am very fortunate to have a good conscience. Otherwise, I would have ended my life – the sooner the better. Hamlet's dreams, which he so feared, held him back from this. No dreams frighten me, despite what Hamlet says. In the light of human suffering, I regard it as no small comfort that a measure of gunpowder hardly costs 4 pennies. It is abominable to go on living when one no longer wants to, but even more dreadful would be undesired immortality. As things stand, however, the whole horrible weight hangs from me by a thread, which I could cut with a penknife.*

(Again aloud to Hamlet) Dreaming can be very consoling; as you know, my Prince, I recently discussed this with Alice (see chapter 2). I find consolation in my science and my dreams. *Some people try to disparage the study of art and dreams, our fantasies, by contending that one is writing books about images. But what are our conversations and our writings if not descriptions of images on our retina or of false images in our mind?*

HAMLET*: You are right, Professor Lichtenberg, *we are such stuff as dreams are made on; and our little life is rounded with a sleep. Sir, I am vex'd; bear with my weakness; my old brain is troubled: be not disturbed with my infirmity.*

LICHTENBERG: All right, Prince Hamlet. But let's return to the important role of unfettered fantasy. *Through aimless drifting, through the aimless wandering of our fantasy we often flush out ideas that studied fantasy can make use of in its orderly household.*

HAMLET: Unfortunately, though, often *a mote it is to trouble the mind's eye. Thus conscience does make cowards of us all, and thus the native hue of resolution is sicklied o'er with the pale cast of thought, and enterprises of great pitch and moment with this regard their currents turn awry and lose the name of action.*

LICHTENBERG: A "philosophical universal formula" would in fact be bland, boring, and utterly meaningless – for linguistic reasons as well. *Most of our expressions are metaphoric; they embody the philosophy of our ancestors. The invention of language preceded philosophy and this is what makes philosophy difficult, especially when one tries to explain it to others who are not practiced thinkers. To express itself, philosophy always has to use the language of nonphilosophy. And that is the way it should be. For each person should be able to express even the most common things in different ways if he wants to follow his own individual feelings; this seldom happens before a certain ripe old age, when one realizes that one is just as good a person as Newton or the village priest or the government official and all our ancestors. Shakespeare provides a taste of this.*

HAMLET: What, this English tabloid author? He supposedly wrote something about me. But we have more important things to discuss today (pathetically, deeply immersed in his thoughts as in the Hamlet film produced by E. Lubitsch in 1942): *To be, or not to be: that is the question: whether 'tis nobler in the mind to suffer the slings and arrows of outrageous fortune, or to take arms against a sea of troubles, and by opposing them end them...*

LICHTENBERG (interrupting Hamlet): What a pathetic and Teutonic preoccupation with principles, Your Royal Highness! *You are letting such things destroy the little peace of mind left to you.*

---

* These lines are actually those of Prospero in Shakespeare's *The Tempest*.

HAMLET: *What should such fellows as I do crawling between heaven and earth?* Can a famous professor like you tell me?

LICHTENBERG: You must provide the answer yourself, Prince

HAMLET: Each person lives on his own island. But I can offer you some help in ferrying around.

Is it a foregone conclusion, then, that our reason is incapable of knowing anything lying beyond reason? Shouldn't man be just as capable of purposefully weaving his ideas of God as the spider weaves its net to trap flies? Shouldn't there exist beings who are just as fascinated by our ideas of God and immortality as we are of the spider and the silkworm?

People have written much about the nature of matter. I wish that matter would for once begin to write about human nature. It would then become apparent that we have still hardly understood each other.

Up to now we have believed that we are the work of something outside us, about which nothing is known or can be known by us other than what the self reveals to us. How would it be, then, if it were the very nature of our being that actually made this world? Here, the earth's orbit and its revolution around its axis are opposed to the orbit of the sun and the hosts of stars around it. Everything is questioned. We fine Christians deplore idolatry – that is, our dear God is not made of wood and gold trim; nonetheless, it still remains an image, which is but one more member of this very series – finer, to be sure, but forever an image. If the spirit wishes to break away from this idolatry, it eventually arrives at the Kantian idea. * *But it is presumptuous to believe that such a mixed being as man will accept all of this in so pure a form. All that the truly wise man can do, then, is to direct everything to a good purpose and yet still accept mankind as it is.* ** (After a short pause) *Sufficient reason to remain silent.*

Hamlet (thoughtful, as in the famous graveyard scene with Yorick's skull):

> *You say, we live on an island?*
> *That physics holds on an island?*
> *In a poor isle, and all of us ourselves*
> *When no man was his own****
> Good, then. *The rest is silence.*

---

* That is, Kant's "categorical imperative."

** Ten days before his death on February 24, 1799, Lichtenberg wrote these lines to his brother.

*** From Shakespeare's island comedy "The Tempest," Act 5, Scene 1, verses 211–212.

What Is Meant by Complex?

Definition of Complexity

In everyday speech we employ the notion of complexity quite naturally: a complex problem, a complex personality, a exceedingly complex task.

In the following discussion, I will try to show that the term "complexity" is useful in describing and understanding highly organized systems when they are classified according to their degree of complexity. This degree of complexity, in turn, has something to do with our ability to describe the system. For instance, a complex machine is more difficult to describe than a simple lever. As the number of parameters required for the complete description of a system increases, so does its complexity.

Complexity may be defined as the logarithm of the number of ways that a system can manifest itself or as the logarithm of the number of possible states of the system: $K = \log N$, where K is the complexity and N is the number of possible, distinguishable states. This definition is borrowed from information theory. The more complex a system, the more information it is capable of carrying[1]. If two systems with M and N possible states are considered, then the number of states for the combined systems is the product $M \times N = \log M + \log N$.

Naturally, this all depends on which possibilities or states one considers or uses in counting. The system represented by the genes and their mutational changes is highly complex. It encompasses the mutational events at the nucleic acids, replication, the DNA repair mechanism, protein biosynthesis, and the interactions of living things within an ecosystem. Accordingly, the number of possible states (N) is very large and the system can manifest itself in a countless number of ways. If, however, one attempts to answer only the question of interest to evolution, whether the newly formed living thing survives in its specific environmental habitat or becomes extinct, then $N = 2$, since there are only two alternatives: survival (1) or extinction (2). Complexity is therefore a relative, operational concept, which must be considered in connection with the system and its function. By changing the question at hand or the functional relationships, one can reduce the degree of complexity. Instead of inquiring into the molecular events of mutation, for example, one might ask only whether the mutants survive.

How are complexity and prediction related, then? A given system has a certain degree of static complexity, which is intrinsic to its structure and effectively constant (an example might be the basic double-helical structure of nucleic acids). Imposed on this static complexity is a dynamical complexity, which changes in time or space. In the system of evolution, this would correspond to the sequence of building blocks in the helical nucleic acids. Prediction is only possible when the initially specified information is completely retained, that is, when the complexity does not change. A prediction becomes less reliable to the extent that the information changes, dissipates, or increases.

Let us consider the problem from the standpoint of information theory. The instructions for the generation of simple, noncomplex systems are short and usually self-evident: for example, add up all the parts and give the sum. Here, complexity can be defined operationally: Seek the smallest program or set of instructions capable of describing a given "structure" – say, a series of numbers. The size of this minimal program, measured in bits, relative to the size of the series of numbers itself is a measure of its complexity.

The series aaaaaaa...is homogeneous (subcomplex). The corresponding program is as follows: write after each a another a. This short program allows the series, no matter how long it is, to be reproduced at will.

The series aabaabaabaab...has a higher complexity, but it is still easy to program: after two a's write b and repeat this operation. Even the series aabaaba*b* baabaaba*b*b... can be described by a very short program: after two a's write b and repeate the operation; at every third repetition, replace the second a by b. Such series have definable structures and the corresponding programs transmit information.

But the series aababbababbbabaaababbab...no longer has a discernible structure. To be programmed, it has to be written out in full. Once the size of a program becomes comparable to that of the system it seeks to describe, the system is no longer programmable. When a structure is no longer describable – that is, when the smallest algorithm necessary to describe it has a comparable number of bits of information to the system itself – I refer to it as fundamentally complex[2]. This definition of fundamental complexity is based on the formulation of A. N. Kolomogorov (1965)[3]. Before this point is reached, however, one encounters quite complicated programs that can still be carried out.

A salient feature of biological structures, which are fundamentally dynamical and hence describable only as "snapshots," is that their description is virtually as complex as the structures themselves. To understand the function of an enzyme, it is necessary to know each coordinate of possibly thousands of atoms at each instant. Almost nothing can be excluded from consideration. Each detail is essential and must be taken into account.

An interesting and still unsolved problem emerges here. It is obvious that a system is indeterminable when its algorithm is just as large as the system itself; limits are somehow set to its practical programmability. But how large is the algorithm permitted to be relative to the structure? Is the system fundamentally complex and hence indeterminable when the difference is just one or a few bits or does the system exhibit this behavior when the difference is greater? The character of fundamental complexity as a criterion of indeterminability seems to imply that it is not possible to establish this limit precisely.

Degrees of Complexity[4] – Fundamental Complexity

A closed system is physically defined as a system that can neither exert forces on the outside nor transfer matter to or accept it from the outside. In such a system all sorts of things can happen; however, provided that no external forces act on it, it remains unchanged in time and space. Nor is self-induced change possible. Münchhausen cannot pull himself out of the swamp by his own queue.

Biological systems constantly interact with their surroundings. They take up substances as nutrients and consume (metabolize) them, they transform chemical energy into heat, they process information. As open systems, they become more and more complex[5]. The structures formed in such systems thus become increasingly complex over the course of evolution. This growth in complexity is associated with an increase in metabolic flux. If this dynamical process of unfolding complexity verges on the limits of structural stability, the system becomes "chaotic." Here, chaotic does not imply the complete lack of structure, but rather the "fundamentally complex" interaction of countless structures in a process of formation and decay, as we have seen earlier. If a bifurcation or branch point is reached, a system becomes fundamentally complex.

This problem has both a *practical* and a *teleological* aspect.

The *practical* aspect: When we ponder and study living phenomena, we seek to abstract our living and thinking personality, but this is possible to only a limited extent – after all, thinking is a phenomenon of human life. To make science possible, all its axioms assume that the thing under consideration can be regarded as an object, that is, as something separate from the observer. For the examination of highly complex biological processes such as evolution or the functioning of the central nervous system, however, this assumption is no longer obviously true. In fact, it no longer holds at all. We cannot distance ourselves from the object being considered; indeed, this is so at the very moment we start to think[6].

Biological processes can attain such a high degree of complexity that it becomes impossible to describe them. The very attempt would be pointless, since the analytical faculties of man, a likewise highly complex being, are insufficient to describe all the parameters of this "person" or "life" in such thorough detail that a "useful prediction" can be made. In this case, the only recourse – as Karl Popper[7] discusses – is to perform "piecemeal technology" instead of "utopian technology." This means that, whereas individual scientific problems can usually be solved piecemeal, a description of a whole of which we ourselves are a part is a utopian fantasy.

The *teleological* aspect: Viewed from the system itself, its complexity is unlikely to cross the dividing line separating it from fundamental complexity, at least not when regulatory mechanisms are available. The system can come to a standstill, a state equivalent to death for living structures. Or it can seek to evolve to a higher stage. Biological evolution, for example, gives rise to new species. Likewise, the central nervous system, forced to respond to an unsolvable problem, either falls under the sway of a sensory illusion such as an optical illusion or perhaps adjusts and thereby "learns" something new. Whatever the response, the system is not destroyed by an increase in the complexity of the demands placed on it.

The complexity of different systems increases at different rates, depending on how many parameters, equations of motion, equations of state, etc., are required to describe the system. At certain stages in this process, the quantitative growth of complexity results in the emergence of qualitatively new properties. Accordingly, a distinction should be made between subcritical, critical, and fundamental complexity.

Subcritical complexity is present when a system, though outwardly complex, can nonetheless be simplified, perhaps mathematically. Simple physical laws, such as Newton's laws, are applicable to the deterministic system thereby obtained. Even in subcritical systems, however, prediction may be difficult in practice – for example, when the method of analysis does not afford sufficient resolution. Without a microscope it is impossible to carry out quantitative bacteriology, even though bacteria grow at a constant rate. Without a computer it is impossible to solve the structure of a macromolecule by X-ray analysis, even though only simple mathematical operations and Fourier transformations are involved. Strictly deterministic initial equations, however, are not a sufficient criterion for subcritical complexity, since these equations can also have chaotic (that is, fundamentally complex) solutions, as we have seen in chapters 4 and 5. Systems exhibiting subcritical complexity are strictly deterministic and allow for exact prediction.

At a certain stage of complexity – its critical value – structures begin to emerge. In the simplest case, convection currents and convection patterns form (chapter 5). I refer to this degree of complexity as *critical complexity.* These systems form subsystems like those encountered in evolution or in irreversible thermodynamics[8]. Their predictability is limited by practical, but not fundamental, considerations. The Heisenberg uncertainty principle belongs to the realm of critical complexity: the exact position and momentum of a subatomic particle cannot be determined simultaneously. This changes nothing in the predictability of events themselves, however. Prediction (foreknowledge) is a macroscopic affair, required of and performed by a macroscopic living thing. Microscopic events are therefore averaged over. In this respect, Pascual Jordan's theory of free will is *wrong*[9]. Free will cannot be regarded as a macroscopic projection of indeterminate microscopic events.

Human free will is to be found instead at an entirely different level: the evolving system of thinking and acting with ever-recurring bifurcations, corresponding to decision-making events. Empowered by his consciousness and legitimized by his special position in the cosmos, his human dignity, and his freedom, man expresses and acts out his will within this system, subject, of course, to the ethical norms of society.

In the central nervous system, the frequency and amplitude modulation (spikes) of electrical activity at the synapses represents structured information, in close analogy to the transmission of information via radio waves. Now and then a neuron may "fire" spontaneously. Although such events reduce the predictability of thinking, they cannot be taken as proof of free will. In fact, the introduction of chance (randomization) is a technique commonly used to obtain exact solutions. By intentional randomization of a deterministic system, a computer can often arrive at an optimal solution. Not seldom, this approach is faster and simpler than calculation of the exact solution. Evolution, the result of random variations, also operates according to this principle. However, since it involves fundamental complexity, it is not predictable. It should be kept in mind, though, that the mere presence of elements of chance does not make a system fundamentally complex.

Systems are *fundamentally complex* if, despite deterministic initial conditions, they have indeterminate or chaotic solutions. In these systems, predictability fails not only for practical reasons but also for fundamental ones. Above all, I include here the integrated or network system called "life," where the whole is more than the sum of the parts. Only by ripping apart the network at some point can we analyze life. We are therefore limited to the study of "dead" things.

The temporal, structural, or energetic transition from critical to fundamental complexity cannot be specified precisely owing to the indeterministic character of fundamentally complex systems. Still, the following objection might be raised: computers, data storage devices, cloud chambers, spectro photometers, and so forth allow indirect monitoring, recording, and control of such integrated systems; therefore, they should compensate for the properties of fundamental complexity and make predictions possible. The reply to this objection is that it is impossible to compensate for the property of fundamental complexity; in the final analysis, a biological organ, our brain, must generate a mental picture in space and time in order to arrive at a prediction. Therefore, the process of monitoring a system has to be conceivable and comprehensible on a human scale and within a human life span.

Eventually, at a certain level of metabolism and associated structure formation, a system verges on a limit of complexity. I refer to this limit as its fundamental complexity. In the evolution of a living system, fundamental complexity often poses a problem. Either the system evades this limit by entering into new relationships ("surmounting" the system, reduction in complexity) or it stagnates.

The following table lists a variety of systems in terms of their degree of complexity. Of course, this table provides only an overview; it is intended to illustrate the "categorical principle of complexity." In addition, the inclusion of numerous "systems" from outside the natural sciences lends a general character to the theory of fundamental complexity. Earlier, I also dealt with phenomena from outside the natural sciences in terms of the scheme chaos-complexity-order. Examples included beauty (chapter 6), the problem of idea and matter (chapter 7), and aging and death (chapter 8). The following key question arises: Is it permissible to apply the theory so broadly? Before addressing myself to this question, however, I would like to discount two possible objections to the theory.

General table

System	Increasing complexity ⟶ Subcritical complexity	Critical complexity	Fundamental complexity
1 Mathematics			
1.1 Axiomatic	Newtonian	Quantum-mechanical (*Planck-Heisenberg*)	*Gödel*
1.2 Programs	None (simple physical law)	Large, but able to master all information	Program comparable in size to the system itself (algorithm = information) (*Chaitin*)
1.3 Models	Differences	Differential equations, solvable in theory, but not always in practice	*Bernoulli* systems, e.g., baker transformation
2 Theory			
2.1 General natural law	Simple law	Statistical law	Law just as large as the experimental data, no theory possible
2.2 Prediction	Not necessary, because simple	Possible in principel, but often not in practice	Not possible
3 Examples from physics			
3.1 Oscillations	Harmonic oscillator, monochromatic vibration	Interference, modulations	Unresolvable vibrational bands
3.2 Hydrodynamics	Heat conduction	*Bénard* patterns	Turbulence
3.3 Statistical physics	Newtonian (Lionvillian)	–	Ergodic
3.4 Physical chemistry (kinetic equations)	Equilibrium Flow equilibrium	Limit cycle Dissipative structures	Chaos

General table (cont.)

System	Increasing complexity → Subcritical complexity	Critical complexity	Fundamental complexity
4 Biology			
4.1 Molecules	Small	Macromolecules	Interactions between macromolecules
4.2 Cells	Cell organelles	Bacteria, amoeba	Not realized biologically, since further biological evolution is blocked. Fundamental complexity cannot be tolerated by the individual: death
4.3 Cellular networks	Aggregates (slime molds)	Multicellular organisms (hydra)	
4.4 Organs	Uniform function	Arranged in the organisms	
4.5 Complex living things	–	Just barely adapted to the ecosystem	
4.6 Nervous system	–	Simple controls, instincts	Central nervous system
5 Evolution			
5.1 Darwinism	Primordial soup	Individual species	Entire biotope
5.2 Replication of nucleic acids	–	Macromolecules with information	Loss of information

General table (cont.)

System	Increasing complexity → Subcritical complexity	Critical complexity	Fundamental complexity
6 Systems outside the sciences			
6.1 Science itself	Phenomenology, Description	Theories, Reproduction	Finalization of science? Destruction of the object (uncertainty relation)
6.2 Philosophy	Simple logic (*Pythagoras*), insights, primal experiences (*Heraclitus*)	Systems	Transcendental philosophy } Transcendence
6.3 Aesthetics	Simple reproduction	Development of style	Art = abstraction of ↗ form → chaos, ↘ content → stagnation
6.4 Language	–	Simple communication, formal language	Speech Poetry
6.5 Religion	"Feelings"	Natural religion, dogmatic religion	Open religion that makes freedom possible "Impossible" religion, e.g., early (and late?) Christianity
6.6 History	Chronicles, anecdotes	Writing of history, historical systems (*Thucydides*, *Hegel*, historicism)	Open history (*K. Popper*)

First objection: "The theory of fundamental complexity is a product of mysticism"

It might be argued that the theory of fundamental complexity contains antinaturalistic, nonscientific, even mystical elements. In this sense, the theory would reflect a certain self-satisfaction "to have investigated everything that can be investigated and to have left what cannot be investigated to contemplation," as Goethe once remarked. This point of view, while understandable, would mean turning one's back on the realm of science.

In the preceding discussion I have shown that the theory necessarily follows from the most recent scientific knowledge about highly complex biological systems. Throughout the macroscopic biological realm, individual molecular events are subject to feedback coupling operating through amplification mechanisms. In this way, statistical fluctuations – that is, deviations from the mean value – can dictate the fate of a system. Under certain conditions networks become indeterminate. Since the theory incorporates the most recent scientific knowledge, it cannot be regarded as either mystical or antinaturalistic. Indeed, a close examination of the axiomatic framework of the natural sciences might reveal that, if the axioms are strictly obeyed, the limits discussed here follow quite naturally. Gödel's theorem (see p. 224), although applicable primarily to mathematics, is equally a proposition limiting the realm of science. Mathematics, too, has to be content with "piecemeal technology."

Second objection: "The theory is a product of scientism."

The second objection might be stated as follows: The theory represents positivism or scientism; in a given field, it seeks and presents proofs that do not belong to the realm of science; it uses materialistic and scientific proofs (the complexity of the central nervous system, the unpredictability of a mutation, etc.) in an area where only logical and philosophical arguments are valid. This objection would be naive. Philosophy and theories of knowledge have always employed the most recent scientific results, interpreted them, and used them for the definition and description of epistemological problems. The ideas of Descartes would have been inconceivable without the heavenly mechanics of Galileo and Kepler, or, better put, he could not have produced his thoughts without them; without Newton's "Principia," David Hume and Immanuel Kant would not have written what they wrote.

It is therefore legitimate and even necessary to interpret the prospects and limitations of science in the light of new scientific knowledge, even when the interpretations have consequences extending to questions of philosophy and the theory of knowledge itself. The narrow line separating the practice of science from scientism would be violated only if the theory were set up primarily to account for, say, our inability to explain or predict history. But this is not its primary intention. Instead, the theory addresses itself mainly to the epistemology of science. Naturally, if it thereby touches the scientific basis of thinking, more far-reaching conclusions are unavoidable. Indeed, it may very well turn out – as discussed in the following section – that the theory also accounts for our inability to explain history in logical terms.

Since neither objection is justified, the theory of fundamental complexity is generally valid.

Zeitgeist and Evolution – Thoughts on the Relation of History and Natural History[10]

"No one can leap ahead of his time, the spirit of his time is also his spirit; rather, it is a matter of knowing the spirit in terms of its content," said Hegel[11], referring to Plato. In his lectures, he said: "It is my desire that this history of philosophy should contain for you a summons to grasp the spirit of time, which is present in us by nature, and – each in his own place – consciously to bring it from its natural condition, i.e. from its lifeless seclusion, into the light of day."[11]

According to this idealistic model of history, "spirit" – Hegel also speaks of a series of true spirits – is the true content of history; it is manifested in human beings and their creations insofar as they apprehend this Zeitgeist and are consciously aware of it. Clearly, this idealistic view of history is no longer valid – at least not in this form – in our current epoch dominated by materialistic and Marxist philosophies of history. Nonetheless, even the most naive observer cannot fail to notice that historical events always seem to behave in violation of the rules. They are describable by neither simple nor complex models. There are no historical laws, but at most rules of probability. This indicates to the observer of historical events that the action of the Zeitgeist, the spirit, or the spirits occurs from outside to regulate or to stir up the world; this influence acts all the more strongly the more man seeks to impose physical order on the world. Ever greater and more ominous is this discrepancy between the technological determinability of our day-to-day world and the seemingly irrational course of human history.

What is responsible for our inability to describe historical events in full and for the seemingly irrational course of history, which is so disquieting to the human spirit? Jürgen Habermas and Niklas Luhmann[12], in their well-known discussion, attempted to reach agreement on a model based on systems analysis. Here, the concept of complexity plays an important role in social (or historical) structures: "Systems maintain and form islands of lower complexity within their stabilizing boundaries..." – in comparison to the largely chaotic surrounding world. It is thus impossible to view the whole of history or even a part thereof. History is largely chaotic and its description is only made possible by observing an isolated system, transferred by the observer, of course, to an isolated region. Only this island of lower complexity is accessible to description. Accordingly, a functioning, viable social system is subject to certain conditions: "...its self-complexity must be sufficient to allow it to respond in a self-preserving way to changes in its surroundings." In other words, the system must have a certain degree of complexity in order to respond in a "living" way. Not at every time and at every place are systems equally well situated or even capable of survival: "The difference in complexity between the world and the system shows that systems are not compatible with every possible surrounding." At the time of the Sun King, the Roi Soleil, the ideas of the French Revolution, though already present in essence, could not be realized; they were not in accord with the Zeitgeist. Moreover, according to Niklas Luhmann (p. 219), "social evolution" represents "an increase in the complexity of society, that is, an increase in the complexity making the society viable in the relation of its individual social systems to the surrounding world."

Leibniz was the first philosopher to consider the complexity of the world. He defined the best of all possible worlds as a world that solved the problem of complexity in an optimal way by combining the greatest possible amount of variety with a minimum of means and abstract hypotheses. His selection criterion for the best of all possible worlds was therefore the "highest complexity for the simplest means" or the "greatest variety for the highest order."

Of course, the notion of complexity is not precisely defined here. But it is regarded as measurable in principle; otherwise, the different worlds could not be assessed according to this criterion of complexity.

Something similar holds true for the Habermas-Luhmann analysis. A quantitative scale of complexity alone is insufficient to explain the formation of social and historical systems or to account for compatibilities and incompatibilities. Then, what is the origin of what I have called, for simplicity's sake, the irrationality of history? Obviously, if we want to use complexity as a key measure in our historical approach, we have to define complexity more precisely; namely, as I stated above, history is fundamentally complex.

We have been introduced to fundamentally complex systems: the system of evolution, the system of the human brain. Of course, these systems are still subject to definite rules, but not the determining and deterministic rules of classical physics. Rather, they are rules of behavior or, as it were, rules of the game. Each of the players in a "systems game" has to obey the rules or else the game would never take place. The rules are as strict as physical laws and, like the rules of chess, they are quite diverse. Nevertheless, the first move is completely open. No or hardly any game of chess proceeds in exactly the same way as another. Not because the rules (corresponding to physical laws) are violated, but because of the fundamentally complex character of each and every game[8].

The human brain possesses the property of fundamental complexity. It tests solutions; it plays with ideas. Although the ways these ideas arise are governed by the "rules" of psychology and neurophysiology, the ideas are nonetheless open, novel, abstruse, dreamy, playful, or creative. The brain is not merely a computer, but also a "random number generator"; it adapts, transforms, and transmits the virtually infinite variety of human environments. The multifarious features of the mind cannot be simulated or "solved" with a computer. Even if a computer had as many elements of circuitry as there are atoms in the entire universe and had performed its calculations since the beginning of the world at the speed of light, it would still have carried out only a finite number of operations – about 10^{120}. The number of different sequences of letters that would fit on a postcard is much larger – even more so the number of possible states of the brain capable of writing or interpreting the postcard.

I therefore propose the following theorem: Because of the fundamentally complex physiological and neurological properties of the brain, there can be no simple causality and no simple predictability in the realm of the mind, in the creation of ideas, in decision-making. In fact, mathematics has reached analogous results in establishing the foundations of decision-making theory (Gödel's Theorem, see p. 224). Its system of mathematical axioms is fundamentally complex. Even in the physical sciences, then, we reach the intrinsic limits of the scientific method when we examine indeterminate systems. A further limitation of the scientific method is the Heisenberg uncertainty principle, which states that both the position and momentum of an elementary particle cannot be precisely measured simultaneously. At issue in the activity of the brain – the body-soul problem – is the consciousness of consciousness. At issue in the quantum-mechanical uncertainty principle is the measurement of measurement. At issue in the propositions about indeterminacy is the logic of logic. The metatheoretical prerequisites of our sciences cannot be firmly established. History is the history of man; it is the sum of the ideas thought by man and the acts committed by man. History is a human brainchild and is manipulated by the human mind. Consequently, a physical viewpoint of history, like that attempted in historicism and Marxism, is untenable. Historicism attempts to describe history in physical and holistic terms in the sense of classical physics and to explain away the difficulties we have mentioned. Karl Popper has dealt with this problem[7] and refuted historicism on philosophical grounds. He argues in favor of an "open" history.

A solid foundation for Popper's statements is provided by the principle of fundamental complexity, since thought, knowledge, and evolution cannot be predicted. As an example of the direct intervention of impassioned thought in documented history, I quote Kleist's description of a spontaneous event that heralded the birth of the French Revolution[13]:

"I am reminded of that thunderbolt which Mirabeau dealt the master of ceremonies who, after the close of the last royal meeting of the king on the 23 of June, at which the king had ordered that the estates be dissolved, returned to the meeting hall where the estates were still gathered and asked them if they had understood the king's command? Yes, answered Mirabeau, we have understood the king's command – it is apparent that he does not really know what he wants. 'But what gives you the privilege' – he continued, and now a spring of awesome ideas spouts within him, to issue commands to us here? We are representatives of the nation. That was just what he needed – 'The nation gives orders and takes none' – to scale the very heights of boldness. 'And so that I make myself absolutely clear to you' – and only at this point did he find the words expressing the entire opposition with which his soul had armed itself – go tell your king that we will not leave our places unless forced to do so at the tip of a bayonet. He then sat back down, self-satisfied, on his chair... One reads that Mirabeau, as soon as the master of ceremonies had left, stood up and suggested that they should constitute themselves (1) as a national assembly and (2) as inviolate."

Is a general theory of historical description at all possible? It is not possible to give a scientific answer to this question. At certain bifurcations or branch points, everything is open. The higher order of idealism thereby made possible – the idealistic view of history – at once penetrates through the materialistic view and leads to the realm of ideas and irrationalities, which, as fundamentally complex systems, are intrinsically unpredictable and thus indeterminate. A metatheoretical foundation for the study of history is not possible. This means, however, that historical structures and theories can exist side by side in a kind of pluralism, with equal validity, in peaceful coexistence. No one historical theory deserves priority.

The Darwinian theory is *the* history of nature. From the Big Bang to Homo sapiens, our world has arisen through the process of evolution. Higher, more complex orders formed, not necessarily through the elimination of earlier orders in a fight to the finish, but rather through the survival of the fittest. Evolution does not mean destruction of what comes earlier; it means change. What we see today as nature represents but a shnapshot (see p. 67). Are we so naive as to believe that everything must remain as it is? But perhaps we have in fact reached the point where biological evolution is at an end, where *we* have brought things to an end, where everything must remain the way it is. Evolution is not only change but extinction. Evolving systems must be capable of dying. Or, conversely, only systems that have "discovered" death can evolve. To ensure its indefinite survival, therefore, a system has to bring evolution to a halt. And this is just the point mankind has reached; it is the source of our crises and our opportunities.

Why did the development of the human brain prove biologically advantageous? We saw in chapter 3 that 4×10^9 bits of information are carried in the double helix of our genetic inheritance, organized as genes, hereditary factors, etc. Mankind now produces roughly 10^{18} bits of new, nongenetic information per year. In other words, the amount of information we produce each year and pass along to the next generation is a billion times greater than what is transmitted through our genetic material over a generation of 30 years. In short, the biological evolution of man has come to an end.

But the end of biological evolution in no way implies that everything has come to a standstill. Biological evolution proceeds according to the Darwinian laws. At the end of the 19th century and well into the 20th century (Lysenko) the adherents of Darwin and those of Lamarck engaged in an acrimonious debate, best illustrated by a simple example. Darwinists gave the following answer to the question "Why do giraffes have long necks?": In the savanna, the ability to feed on the leaves of high trees was advantageous for survival. Certain long-necked antelopes were therefore better able to survive; they multiplied, while others became extinct. This process, operating over millions of years, gradually led to the development of giraffes. Lamarckists replied: Giraffes constantly stretched their necks to reach the leaves. They thereby "trained" their necks and were then able to pass along the acquired trait of "a long neck" directly to their offspring.

Amazingly enough, the human brain has transformed man into a Lamarckian creature! If he wishes to fly, he does not have to wait thousands of years until he develops wings. He invents flying machines instead. If he does not want to freeze, he does not have to inherit a polar-bear fur coat passed along genetically over thousand of years. Either he shoots a polar bear and uses its fur coat or he turns up the central heating. We, as Lamarckian creatures, are able to turn our wishes directly into genetic realities, into "genetic traits." Yes, genetic traits! Though not those recorded in the double helix, nonetheless they represent knowledge, abilities, instruments, customs, traditions, and moral prescriptions; they are set down and passed on through education, libraries, colloquial speech, social norms, and political systems.

What is qualitatively new in the current situation is that, over the last few years, man has learned to apply these technical and manipulative abilities to his own nature as well as to his physical surroundings. He can impose his ideas on nature. Natural history has become history[14].

What does this mean? Ever since prehistoric times when man began to intervene in nature, he has influenced the evolution of some parts of nature – often not intentionally, but rather through symbiosis. To give just one example: For the jackals lurking around the fire pits of ancient man in the hope of snatching a scrap of food, it became more advantageous to become a domestic dog. As long as man was a part of nature, this behavior merely accelerated the natural development in a particular direction; it was still a part of natural evolution and therefore not within man's responsibility. Modern man, too, is a biological creature and thus a part of nature. As a result of his activities, however, biological evolution has been overtaken by technological and cultural progress. That is, evolution now lies in the hands of man; it is *his* responsibility.

The industrial age, beginning roughly one-hundred and fifty years ago, and especially the biotechnology age, beginning little more than a decade ago, have resulted in an unprecedented interaction between the realm of ideas (Popper's World III) and the realm of nature (the realm of evolution). This novel feedback coupling brought about by man and for which he is responsible is capable of imposing on natural history the same instability, the same degree of complexity, the same tendency toward crisis that we observe in history. This interaction threatens to grow out of control, leading to a global ecological catastrophe or to atomic destruction or to total genetic manipulation. Although man stands in the midst of history, he makes his *own* history. The forms of matter, the biological fundaments, and the material realities of this world governed life *and* thought in the preindustrial age. But they have become increasingly circumstantial and are being pushed aside by scientific advances in favor of a history shaped exclusively by the dominion of human thought.

Human responses are dictated by the responses of the central nervous system, including the hormonal system under its control. They are infinitely diverse. For this reason history is qualitatively different from natural history. By conquest and rapacious exploitation of nature, man has not only come to dominate nature but also to impose on it his historical laws and instabilities. Natural history or evolution – which only now, at the end of the 20th century, is becoming fully manifest to us – *could* be regarded as complete. From a historicist viewpoint, we could look back with satisfaction, regarding the passing epoch as a barbaric period during which the cruel and unflinching laws of evolution operated and from which, at last, we are now liberated after ten million years of human history. *We could, but we should not!*

At this crossroads of history and natural history – for the first time in the history of our world – *we* have been given the moral responsibility not only for our *history* (and its crimes) but also for *natural history* (and the crimes committed against it). And we will have to bear this cross[15].

Departure from Intrinsic Truths – On the Impossibility of Proving Proofs

Man is a fundamentally complex creature and he is mortal; the two are related (see chapter 8). Thus, once and for all, he is self-referential; this also means that he cannot lay claim to completeness.

But physics and mathematical logic are also self-referential; they are also self-coupled by feedback. Kurt Gödel formulated this in his famous statement: "All non-self-contradictory axiomatic formulations of number theory contain undecidable propositions."[16] Douglas R. Hofstadter has explained this in more detail:[17] "The proof of Gödel's Incompleteness Theorem hinges upon the writing of a self-referential mathematical statement, in the same way as the Epimenides paradox is a self-referential statement of language. (The Epimenides paradox is the statement: A Cretan says, "All Cretans are liars.") But whereas it is very simple to talk about language in language, it is not at all easy to see how a statement about numbers can talk about itself. In fact, it took genius merely to connect the idea of self-referential statements with number theory. Once Gödel had the intuition that such a statement could be created, he was over the major hurdle. The actual creation of the statement was the working out of this once beautiful spark of intuition...

Gödel had the insight that a statement of number theory could be *about* a statement of number theory (possibly even itself), if only numbers could somehow stand for statements. The idea of a *code*, in other words, is at the heart of his construction. In the Gödel code, usually called Gödel-numbering, numbers are made to stand for symbols and sequences of symbols. That way, each statement of number theory, being a sequence of specialized symbols, acquires a Gödel number, something like a telephone number or license plate, by which it can be referred to. And this coding trick enables statements of number theory to be understood on two different levels: as statements of number theory, and also as *statements about statements* of number theory.

Once Gödel had invented this coding scheme, he had to work out in detail a way of transporting the Epimenides paradox into a number-theoretical formalism. His final transplant of Epimenides did not say, This statement of number theory is false, but rather, This statement of number theory does not have any proof. A great deal of confusion can be caused by this, because people generally understand the notion of 'proof' rather vaguely. In fact, Gödel's work was part of a long attempt by mathematicians to explicate for themselves what proofs are. The important thing to keep in mind is that proofs are demonstrations *within fixed systems* of propositions. In the case of Gödel's work, the fixed system of number-theoretical reasoning to which the word 'proof' refers is that of *Principia Mathematica (P.M.)*, a giant opus by Bertrand Russell and Alfred North Whitehead, published between 1910 and 1913. Therefore, the Gödel sentence G should more properly be written in English as:

> This statement of number theory does not have any proof in the system of *Principia Mathematica.*

Incidentally, this Gödel sentence G is not Gödel's Theorem – no more than the Epimenides sentence is the observation that 'The Epimenides sentence is a paradox.' We can now state what the effect of discovering G is. Whereas the Epimenides statement creates a paradox since it is neither true nor false, the Gödel sentence G is unprovable (within *P.M.*) but true. The grand conclusion? That the system of *Principia Mathematica* is 'incomplete' – there are true statements of number theory which its methods of proof are too weak to demonstrate.

But if *Principia Mathematica* was the first victim of this stroke, it was certainly not the last! The phrase 'and Related Systems' in the title of Gödel's article is a telling one; for if Gödel's result had merely pointed out a defect in the work of Russell and Whitehead, then others could have been inspired to improve upon *P.M.* and to outwit Gödel's Theorem. But this was not possible: Gödel's proof pertained to *any* axiomatic system which purported to achieve the aims which Russell and Whitehead had set for themselves. And for each different system, one basic method did the trick. In short, Gödel showed that provability is a weaker notion than truth, no matter what axiomatic system is involved.

Therefore Gödel's Theorem had an electrifying effect upon logicians, mathematicians, and philosophers interested in the foundations of mathematics, for it showed that no fixed system, no matter how complicated, could represent the complexity of the whole numbers: 0, 1, 2, 3,.. Modern readers may not be as nonplussed by this as readers of 1931 were, since in the interim our culture has absorbed Gödel's Theorem, along with the conceptual revolutions of relativity and quantum mechanics, and their philosophically disorienting messages have reached the public, even if cushioned by several layers of translation (and usually obfuscation). There is a general mood of expectation, these days, of 'limitative' results – but back in 1931, this came as a bolt from the blue.

Gödel's discovery meant the impossibility of absolute solutions, the departure from intrinsic truths. Man's thinking and being are self-referential, self-coupled by feedback – as I said above. Accordingly, man cannot exceed his limitations; he is incapable of transcendence. Does this imply resignation? I do not think so, though it could very well lead to skepticism toward hasty conclusions and actions.

Odo Marquard has written[18]: "Skepticism desires the elusive individual, the total individuality. But it expects the unavoidable individual, that is, each man, because he must die 'unjustifiably' and is mortal. Therefore, the life of man is always too short to allow him to change and thereby free himself, to the extent he chooses, from what he already is: he simply does not have sufficient time for this. He must always remain, therefore, what historically he mainly was: he has to 'carry on.' A future requires a past: what I have chosen to be is conveyed by what I have not chosen to be, and the latter is always so overriding for us that – owing to our short life span – it exceeds our capacity for rationalization. For this reason, if one wants to rationalize at all – under the time pressures of our vita brevis – it is necessary to rationalize what has been chosen (the change), rather than what has not been chosen: he who changes bears the burden of proof. By assuming this rule, which follows from human mortality, skepticism tends toward conservatism. 'Conservative' is here a totally unemphatic term, the explanation of which is best left to surgeons. They consider whether a 'conservative' treatment is called for, or whether the kidneys, a tooth, an arm, or part of the intestines should be removed: one cuts lege artis only when one has to (when there are compelling indications), otherwise not an all, and never everything; there is no operation without conservative treatment; for one cannot cut the entire person out of a person. Whether intentional or not, this is overlooked by those who bristle at the notion of conservatism. In the same way, not everything is subject to change and hence not every lack of change is open to criticism: Therefore, those who do this in the sense of an 'overtribunalization of reality' – from the philosophers of history to the philosophers of discourse – attain something different from what they intended. The overtribunalizers do not achieve an absolute rationality, but rather an 'escape into unaccountability,' which advocates freedoms that we – prior to all intrinsic rights – already are; to these belong commonplaces. Because we die too quickly for total changes and total rationalizations, we need commonplaces, including the commonplace that is philosophy. The skeptic therefore accepts traditions as unavoidable owing to our mortality; and what is known there – commonly and with the status of commonplaces – they know as well. Skeptics are therefore not at all those who intrinsically know nothing; they merely know nothing intrinsic: skepticism is not the apotheosis of uncertainty, but rather the departure from what is intrinsically true."

Our ways of thinking and our logical systems have hierarchical structures. If a system is too complex to be manageable – for example, a central nervous system in evolution – then it breaks apart into subsystems and thereby reduces its complexity. In the study of neurophysiology, the nerve cell is treated as a functional unit that generates an electrical potential and 'fires,' resulting in, say, muscular contraction. Only the firing of the nerve fibers is of interest to physiologists interested in the system of muscular contraction, not how this 'firing' comes about or what biochemical processes are involved. They reduce the complexity to what is operationally essential. Of interest to cell biologists, on the other hand, are the cell-biological processes occurring in the cell membrane, which they seek to elucidate through specific investigations; they thereby attempt to reduce the complexity of the system. They are interested hardly at all – at least, not in connection with these studies – in the genetics of nerve cells, which, in turn, are of primary interest to molecular biologists, and so forth. Reduction of complexity is a natural approach to setting constraints on fundamental complexity. Where exactly the actual threshold of knowledge lies is hard to determine and is itself reduced to an operational problem.

Plato illustrated this in his cave analogy. Imagine cave dwellers who from birth have never known anything other than the interior of their cave. Since they cannot even stand at the entrance and gaze at the outside world, they have no knowledge of it. Only occasionally does a shadow pass along the wall near the entrance to the cave. The cave dwellers, perceiving a fuzzy silhouette and hearing indistinct noises, form mythical notions of what is occurring outside the cave and eventually develop an entire mythology based on these notions. By chance one of the cave dwellers manages to reach the outside world and to explore it. When he finally returns to the cave after a long period of time and reports on his experiences – sun and rain, men and animals, houses and rivers – he is ridiculed as a dreamer and a lunatic by the other cave dwellers, who are able to live only with their reduced level of complexity.

Some Consequences for Our Society and for Future Research

Freedom and Arbitrariness – Trust Is a Principle After All

Because of the fundamentally complex nature of all human thought and action – the feedback character of all systems with which we are confronted, including ourselves, the uncertain and indeterminable transition between order and chaos as characterized by the Feigenbaum number, the network character of our central nervous system and of life in general, there always remains an indeterminable element in what we do, in our thought and action. This indeterminable element does not constitute our freedom, however. If this were so, then freedom would be due to nothing more than statistical fluctuation. It would be the freedom of a water flea.

Human freedom has its origin in the cultural sphere; man has freed himself, as Friedrich von Hayek[19] says: "Freedom became possible through the stepwise development of the discipline of civilization, which is, at the same time, the discipline of freedom. By means of impersonal and abstract rules, it protects people from the arbitrary force of others and enables the individual to create for himself a sphere in which no other person is permitted to interfere and within which he can use his knowledge for his own purposes."

Freedom results in an enormous enhancement of complexity. In the case of boundless individualism, however, it can lead to the total destruction of society, thereby pulling the rug from under freedom itself. Without a society in which he is integrated and which exerts some control over him, no man is free. Rather, he is a poor, harried creature of nature, without the advantages arising from division of labor and without protection, left alone to face annihilation by the physical realities. The enormous increase in complexity due to freedom, on the other hand, is useful to society – a freely evolving society is more efficient than a regimented society. Yet the complexity must still be reduced in some way. This is accomplished in human societies through certain conventions based on mutual trust.

Without a modicum of trust, a free life cannot be mastered. Each morning, at the moment of our awakening, we would be overcome by the fear of what might happen during the coming day and would probably pull the blanket over our head. Without trust there would be no commerce; we rely on others to pay us what they owe us. Without trust there would be no contracts; we expect agreements to be fulfilled. Without trust there would be no vehicular traffic; we expect others to obey traffic rules. Naturally, trust is violated on occasion through illnesses, tricksters, accidents. Nonetheless, a free society cannot exist without trust.

Niklas Luhmann views trust as a reduction of complexity:[20] "We can now regard the problem of trust as a problem of a risky advance. The world has drawn asunder into uncontrollable complexities; hence, at any given moment, other people are capable of choosing totally different ways of acting. But I have to act here and now. The moment when I am able to see what others do and, seeing this, to adjust accordingly, is short. In it alone, there is little complexity to grasp and work out and therefore little rationality to gain. More complex realities afford more chances, if I am willing to trust that others will act in a certain way in the future."

Trust is thus a principle of human coexistence, but it is not possible without love. Love, therefore, is a necessary principle in our human world. By no means do I wish to imply that love is merely a product of developmental biology and evolution. Love is a power that acts in our world. Without this power the human world could not exist.

Evolutionary Biology as an Object of Research

The questions posed by evolutionary biology are and will continue to be scientific problems of the utmost import. However, a total description of the system of evolution, be it mechanical or holistic, is beyond us. Scientific questions necessarily focus on specific problems, such as the structure of the ribosome or the deciphering of the genetic code. These problems, in turn, must be of unquestionable value in advancing our state of knowledge and, possibly, of practical importance as well. Otherwise, they remain so general and mathematical in nature that they barely touch the real diversity of biological structures. This diversity constitutes the whole, which is fundamentally complex.

A further consequence of this realization is the impossibility of using genetic manipulation, based on scientific extrapolation of the laws of evolution, to create a complete plant, animal, or human being. Genetic manipulation will always remain "piecemeal technology"; perhaps it would be better to refer to it as "amateur technology."

Our Brain – A Complex Organ for the Reduction of Complexity

Since the very beginning of recorded thought, man has been inclined to ponder the relationship of body to mind, of body to soul. Unquestionably, the human mind has its domicile in the body, even if one assumes the existence of a spiritual realm, the sphere of ideas, metaphysics as qualities separated from the body. At the very least, the body is the fundament supporting the mind. There is much feedback between body and mind. We usually think better when we possess a healthy physical constitution. Conversely, thoughts enable us to rise to heights of physical accomplishment.

Psychological pain and unassimilated experiences, on the other hand, sometimes hurt our body. What, then, is the relationship of body to mind?

The organ we use to think with is called the brain. This complicated organ, containing billions of nerve cells interconnected with one another in countless ways, possesses an unimaginable level of complexity. How does such a complex quantity of objects give rise to a new quality, human thought? Will future advances in research bring us closer to a description of the human mind in physical terms?

The underlying structure of the brain is genetically programmed: its size, the number of neurons, the rough assignment of specific functions (hearing, vision, etc.) to certain regions. It also bears the imprint of genetically inherited programs governing some processes of thought and behavior. But the brain is also capable of learning. In a certain sense, it is an intelligent computer, one that is able to learn. Specific synaptic connections are formed during the course of the first year of human life. Wires (nerve fibers) grow toward specific biological targets, including other nerve fibers; they form countless connections and gradually transform the entire system into an intricate network of interconnections. The more practice the nerve functions get, the better these connections become; more of them are made and they are formed faster.

The central nervous system, then, is doubly complex. On the one hand, its structure and function are governed by many hundreds of genes functioning simultaneously in an integrated fashion. On the other hand, many key abilities are epigenetic; they must be developed. The basic structure of the brain arose over the course of evolution through genetic adaptation to the physical environment. During the life of the individual, however, the brain can develop further through epigenetic adaptation to the individual's surroundings. Even a seemingly simple concept like "intelligence," therefore, is impossible to define. And what cannot be defined cannot be modified in a controlled fashion.

There are two fundamentally different points of view in neurophysiology and much effort is currently being spent in trying to unify them. Systems physiologists regard the brain and human behavior from a *generalizing* standpoint. Neurochemists maintain a *molecular-biological* standpoint.

The generalizing standpoint of systems physiologists presupposes that the system cannot be understood as the sum of its separate parts. Therefore, collective phenomena – say, neuronal processing of visual images along the pathway from the retina to the cerebral cortex – are considered and individual phenomena intentionally ignored. This direction of research will therefore always remain phenomenology, albeit a very refined form thereof with wide-ranging consequences.

The molecular-biological standpoint presupposes that molecular events govern the overall biological processes. This is a correct and necessary assumption. However, the central nervous system is a fundamentally complex network in which individual chemical events lead to macroscopic processes in a unpredictable way. There is no such thing as an individual nerve cell as a basic element of thought. Therefore, this approach, too, is forced to use simple models to explain certain phenomena, including possibly pathological phenomena such as mental illness. Any attempt, however, to decipher the program of the central nervous system in order to make absolutely reliable predictions is doomed to failure.

An atomistic (Democritean) point of view is not necessarily the only path open to science. Molecular biology has made great strides along the atomistic path and can lay claim to much success. The realization that the human brain cannot be described completely by this approach does not represent a failure in the underlying method, but rather points to the inherent limitations in its axiomatic structure (including the principle of fundamental complexity). The standpoint of systems physiologists aims at unraveling the course of the overall system. These approaches might be illustrated by way of a railroad system. A railroad schedule does not represent the reality of the railroad network, but it does provide an exact description adapted to a specific purpose – in this case, to inform travelers how to arrive at a certain destination. For one group of travelers, those traveling between major cities, the complete schedule is not even necessary. They require only the schedule of trains traveling between their point of origin and the city in question. This schedule is an extract of an extract, as it were. The presence of switches, of switch towers, of signals that have to be set properly, of people to clean the signal lamps, of station attendants and of people to prepare the schedules – this is of little interest to these travelers. They reduce the complexity of the railroad system to what they need to know, to what is knowable to them.

At the Boundaries of Science

Science is a product of the human mind. As a product, it cannot surpass its producer. Philosophically, this has been known at least since Kant and was reconfirmed in a different way by Karl Popper. It is now established that the realm of biology, particularly the realm governed by the central nervous system, possesses the property of fundamental complexity.

How should one proceed, then, at the boundaries of science? The answer to this question is difficult, multifaceted, and individual. The Soviet mathematician Igor R. Shafarevitch[21] had the following to say in 1973 when he received the Heinemann Prize in Göttingen: "Without a definite goal mathematics can develop no idea of its own form. All that would remain as an ideal would be an uncontrolled growth or, better put, an expansion in all directions. To make a different comparison, one could say that the development of mathematics differs from the growth of a living thing, which preserves the form imposed upon it. The development in the former case more resembles the growth of a crystal or the diffusion of a gas, which expands freely until it encounters an external barrier."

And he comes to the conclusion: "The goal of mathematics cannot lie in a low-level kind of human endeavor, but rather in a high-level one, in religion. Now, it is naturally very difficult to imagine how that can come about. But it is even more difficult to imagine how mathematics can continue to develop indefinitely without some knowledge of a deeper meaning."

We Island Dwellers – The Beautiful Life on the Archipelago

The circle closes. We have reached the boundaries of our science and cannot go further. "For now we see through a glass, darkly; but then face to face: now I know in part; but then shall I know even as also I am known."[22]

Our world is an island, an island of order, an island of physical laws, an island of ideas, an island of trust. We live on our island. *Since the world is harmonic* (chapter 6), *ideas and matter exist in a creative relationship with each other* (chapter 7), *the world is finite* (chapter 8), *love is a strong, integrating power in it* (chapter 9), *we live in a beautiful world!*

Perhaps there are other islands – maybe even an entire archipelago. Their order could be different. If it is, we must accept it as equal to ours, since we now recognize the plurality of this world.

"We have (...) not only traveled through the land of pure reason and gazed carefully at every part thereof, but we have also measured it and determined the position of everything. This land is an island, however, and is enclosed in unchangeable boundaries by nature itself. It is the land of truth (a charming word), surrounded by a wide and stormy ocean, the realm of appearance, where many a fog bank and much ice, soon to melt, gives the appearance of new land. It repeatedly deludes the seafarer in search of discovery with empty hopes, leads him on to adventure, from which he never desists and which are never-ending."[23]

Friedrich Hölderlin[24]

Mnemosyne III

Ripe are, dipped in fire, cooked
The fruits and tried on the earth, and it is a law,
Prophetic, that all must enter in
Like serpents, dreaming on
The mounds of heaven. And much
As on the shoulders a
Load of logs must be
Retained. But evil are
The paths, for crookedly
Like horses go the imprisoned
Elements and ancient laws
Of the earth. And always
There is a yearning that seeks the unbound. But much
Must be retained. And loyalty is needed.
Forward, however, and back we will
Not look. Be lulled and rocked as
On a swaying skiff of the sea.
. . .

References

1. Life – A Dynamical System between Order and Decay

1 W. d'Arcy Thompson: On Growth and Form. Cambridge Univ. Press, Cambridge 1952.

2 J. Lisziewicz, A. Godany, H.-H. Förster, H. Küntzel: Isolation und Nucleotide Sequence of a Saccharomyces cerevisiae Protein Kinase Gene Suppressing the Cell Cycle Start Mutation cdc25. J. Biol. Chem. *262* (1987) pp. 2549–2553.

3 René Descartes: Discours de la méthode pour bien conduire sa raison et chercher la vérité dans les sciences, Paris 1637.

4 P. Glansdorff, I. Prigogine: Thermodynamic Theory of Structure, Stability and Fluctuations. New York 1971.

5 S. C. Müller, T. H. Plessner, B. Hess: The Structure of the Core of the Spiral Wave in the Belousov-Zhabotinskii-Reaction. Science *230* (1985) pp. 661–663.

6 B. Hess, A. Boiteux: Oscillations in Biochemical Systems. Ber. Bunsenges. Phys. Chem. *84* (1980) pp. 392–398.

7 M. Markus, B. Hess: Transitions between oscillators modes in a glycotytic model System. Proc. Natl. Acad. Sci. USA *81* (1984) pp. 4394–4398.

8 A. Boiteux, A. Goldbeter, B. Hess: Control of Oscillating Glycolysis of Yeast by Stochasticy, Periodic and Steady Source of Substrate: A Model and Experimental Study. Proc. Natl. Acad. Sci. USA *72* (1975) pp. 3829–3833.

9 James D. Watson: The Double Helix. Atheneum, New York 1968.

10 Translated by Janette C. Hudson from Marianne Burkhard: Conrad Ferdinand Meyer, Twayne Publishers, Boston 1978, pp. 99.

2. Biochemistry – Gain through Chaos

1 Lewis Carroll: Alice's Adventures in Wonderland, Penguin Books, Harmondsworth, Middlesex 1946.

2 Emil Fischer: Einfluß der Konfiguration auf die Wirkung der Enzyme. Berichte der Gesellschaft Deutscher Chemiker *27* (1894) p. 2985.

3 Friedrich Cramer: Einschlußverbindungen. Heidelberg 1954.

4 N. Hennrich, F. Cramer: Inclusion compounds, XVIII The catalysis of the fission of pyrophosphates by cyclodextrin. A model reaction for the mechanism of enzymes. J. Am. Chem. Soc. *87* (1965) pp. 1121–1126; F. Cramer, W. Saenger, H. C. Spatz: Inclusion com-

pounds, XIX The formation of inclusion compounds of α-cyclodextrin in aqueous solutions. Thermodynamics and kinetics. J. Am. Chem. Soc. *89* (1967) pp. 1–20.

5 J.M. Lehn: Supramolecular Chemistry – Scope and Perspectives. Molecules, Supermolecules, and Molecular Devices (Nobel Lecture). Angew. Chem. Int. Ed. Engl. *27* (1988) pp. 89–112; Angew. Chemie *100* (1988) pp. 91–116.

6 F.v.d. Haar, F. Cramer: Hydrolytic Action of Aminoacyl-tRNA Synthetases from Baker's Yeast: "Chemical Proofreading" Preventing Acylation of $tRNA^{Ile}$ with Misactivated Valine. Biochemistry *15* (1976) pp. 4131–4138; W. Freist, I. Pardowitz, F. Cramer: Isoleucyl-tRNA Synthetase from Baker's Yeast: Multistep Proofreading in Discrimination between Isoleucine and Valine with Modulated Accuracy, a Scheme for Molecular Recognition by Energy Dissipation. Biochemistry *24* (1985) pp. 7014–7023. F. Cramer, W. Freist: Molecular Recognition by Energy Dissipation, a New Enzymatic Principle: The Example Isoleucine-Valine. Accounts of Chemical Research *20* (1987) pp. 79–84.

7 A. Ansari, J. Berendzen, S.F. Bowne, H. Frauenfelder, I.E.T. Iben, T.B. Sauke, E. Shyamsunder, R.D. Young: Protein States and Proteinquakes. Proc. Natl. Acad. Sci. USA *82* (1985) pp. 5000–5004.

8 J.D. Bleil, P.M.W. Assarman: Mammalian sperm-egg interaction: Identification of a glycoprotein in mouse egg zonae pellucidae possessing receptor activity for sperm. Cell *20* (1980) pp. 873–882.

9 F. Cramer, H.-J. Gabius: New Carbohydrate Binding Proteins (Lectins) in Human Cancer Cells and their Possible Role in Cell Differentiation and Metastasation; B. Pullman et al. (Eds.): Interrelationship Among Ageing, Cancer and Differentiation. Dordrecht 1985, pp. 187–205.

10 H.-J. Gabius, R. Engelhardt, F. Cramer: Expression of Endogenous Lectins in Human Small-Cell Carcinoma and Undifferentiated Carcinoma of the Lung, Carbohydrate Res. *164* (1987) pp. 33–41.

11 F. Cramer, H.-J. Gabius: Tumorspezifische Lektine und ihre Rolle bei der Metastasierung. Tumor-Diagnostik und Chemotherapie mit Hilfe tumorspezifischer Lektine. Jahrbuch Akad. Wiss. Göttingen 1986, pp. 143–145.

12 H.-J. Gabius, R. Engelhardt, G. Graupner, F. Cramer: Lectin in Carcinoma Cells: Level Reduction as Possible Regulatory Event in Tumor Growth and Colonization, in: Lectins, Vol. V., T.C. Bøg-Hansen, E. van Driessche (Eds.). Berlin/New York 1986, pp. 237–242.

13 H.-J. Gabius, K. Vehmeyer, R. Engelhardt, G.A. Nagel, F. Cramer: Carbohydrate-Binding Proteins of Tumor Lines with Different Growth Properties: Changes in Their Pattern in Clones of Transformed Rat Fibroblasts of Differing Metastatic Potential. Cell Tissue Res. *246* (1986) pp. 515–521.

14 P. Pfeifer, U. Welz, H. Wippermann: Fractal Surface Dimension of Proteins: Lysozyme. Chem. Phys. Lett. *113* (1985) pp. 535–540.

15 A.C. Wilson, S.S. Carlson. T.J. Lighte: Biochemical Evolution. Ann. Rev. Biochem. *46* (1977) pp. 437–639.

16 M. Eigen, E. Schuster: The Hypercycle. A Principle of Natural Self-Organization. Berlin/Heidelberg/New York 1979 (see also Chap. 4).

17 Taken from: The Oxford Book of English Verse, 2nd ed., Oxford Univesity Press, London 1939, p. 966.

3. Genes, Genetic Maps, Gene Therapy – A Problem of Complexity

1 The original Goethe quotes are taken from J.P. Eckermann: Gespräche mit Goethe, 2 volumes, Frankfurt 1981; The Darwin quotes are taken from Charles Darwin: On the Origin of Species by Means of Natural Selection or the Preservation of Favored Races in the Struggle for Life, John Murray, London 1859; See also Friedrich Cramer: Denn nur also beschränkt war ja das Vollkommene möglich ... Eine wissenschafts- theoretische Interpretation von Goethes Gedicht "Metamorphose der Tiere". Sitzungsbericht der Heidelberger Akademie der Wissenschaften, Math.-Nat. Klasse. Heidelberg 1983, pp. 17–30.

2 For further details of the molecular biology, see H.G. Gassen, A. Martin, G. Sachse: Der Stoff aus dem die Gene sind. München 1986.

3 James D. Watson: Molecular Biology of the Gene. New York 1977, p. 178.

4 A.M. Maxam, W. Gilbert: A New Method for Sequencing DNA. Proc. Natl. Acad. Sci. USA *74* (1977) pp. 560–564.

5 F. Sanger, S. Nickler, A. Coulsen: DNA Sequencing with Chain Terminating Inhibitors. Proc. Natl. Acad. Sci. USA *74* (1977) pp. 5463–5467.

6 Taken from the laboratory notebook of Dr. Sabine Englisch, Max-Planck-Institut für experimentelle Medizin, Göttingen.

7 U. Englisch, S. Englisch, P. Markmeyer, J. Schischkoff, H. Sternbach, H. Kratzin, F. Cramer: Structure of the Yeast Isoleucyl-tRNA Synthetase Gene (ILS 1). DNA-Sequence, Amino-Acid Sequence of Proteolytic Peptides of the Enzyme and Comparison of the Structure to those of other Known Aminoacyl-tRNA Synthetases. Biol. Chem. Hoppe-Seyler *368* (1987) pp. 971–979.

8 J.C. Catlin, F. Cramer: Desoxyoligonucleotide Synthesis via the Triester Method. J. Org. Chem. *38* (1973) pp. 245–250; see also E.L. Winnacker: From Genes to Clones. Introduction to Gene Technology. VCH, Weinheim 1987; Gene und Klone. Verlag Chemie, Weinheim 1984, pp. 44–59.

9 A. Prouska, T.M. Pohl, D.P. Barlow, A.M. Frischauf, H. Lehrach: Contruction and use of human chromosome jumping libraries from Notl digested DNA. Nature *325* (1987) p. 353.

10 Peter Koslowski, Philipp H. Kreuzer, Reinhard Löw (Eds.): Die Verführung durch das Machbare. Ethische Konflikte in der modernen Medizin und Biologie. Stuttgart 1983.

11 Reinhard Löw: Leben aus dem Labor. Gentechnologie und Verantwortung – Biologie und Moral. München 1985.

12 Reinhard Löw: Philosophie des Lebendigen. Frankfurt/M. 1980.

13 Wolfgang van den Daele: Menschen nach Maß? Ethische Probleme der Genmanipulation und Gentherapie. München 1985.

14 Friedrich Cramer: On the Necessity of Guidelines for Genetic Research and the Impredictability of Biological Sciences, in: E. Mendelsohn, D. Nelkin und P. Weingart (Eds.): The Social Assessment of Science, Proceedings, 13th Science Studies Report of the University of Bielefeld. 1978, pp. 36–56.

15 Friedrich Cramer: Die neue Biologie – Akzeleration oder Perversion der natürlichen Evolution? Tijdschritt voor de Studie von de Verlichting en het vrije Denken *8/9* (1980/81) pp. 159–169.

16 Friedrich Cramer: Fundamental Complexity. A Concept in Biological Sciences and Beyond. Interdisciplinary Science Reviews *4* (1979) pp. 132–139.

17 Friedrich Cramer (Ed.): Forscher zwischen Wissen und Gewissen. Berlin/Heidelberg/New York 1974.
18 Friedrich Cramer: Gibt es wissenschaftliche Tabus? Zeitwende *51* (1980) pp. 136–145.
19 Friedrich Cramer: Gene: Gefahr und Gewinn. Sind wir auf dem Wege zum biologischen Unmenschen? Freibeuter *9* (1981) pp. 71–78.
20 Wolfgang Hildesheimer: Mozart. Frankfurt 1977.
21 Jürgen Habermas: Erkenntnis und Interesse. Frankfurt/M. 1975.
22 Friedrich Cramer: Fortschritt durch Verzicht. München 1975.
23 Takcn from: Thc Norton Anthology of American Literature, Vol. 2, 2nd ed., W. W. Norton and Company, New York, London, 1985, p. 1148.

4. Evolution – Phylogenetic Trees and Lightning

1 The original Einstein quotes are taken from: Albert Einstein – Max Born. Briefwechsel 1916–1955. München 1969.
2 Charles Darwin: On the Origin of Species by Means of Natural Selection or the Preservation of Favored Races in the Struggle for Life, John Murray, London 1859.
3 Thomas Kuhn: The Structure of Scientific Revolutions, University of Chicago Press, Chicago 1962.
4 H. Küntzel, B. Piechulla, U. Hahn: Consensus Structure and Evolution of 5srRNA. Nucleid Acids Res. *11* (1983) pp. 893–900.
5 F. Ayala: The Mechanism of Evolution: Scient. Amer. *239* (Sept. 1978) p. 48.
6 Jacques Monod: Chance and Necessity, Knopf, New York 1971.
7 Manfred Eigen, Peter Schuster: The Hypercycle – a Principle of Natural Self-Organization. Heidelberg/ New York 1979.
8 S. Spiegelman: An Approach to the Experimental Analysis of Precellular Evolution. Quarterly Reviews Biophysics *4* (1971) pp. 213–253; I. Haruna, S. Spiegelman: Specific Template Requirements of RNA Replicases. Proc. Natl. Acad. Sci. USA *54* (1965) pp. 579–587.
9 P. Schuster: Evolution von Molekülen zu Gesellschaften. Physik in unserer Zeit *14* (1983) pp. 66–80.
10 P. Schuster: Polynucleotide Replication and Biological Evolution, in: E. Frehland (Ed.): Synergetics: From Microscopic to Macroscopic Order. Berlin/Heidelberg/ New York/Tokio 1984, pp. 106–121.
11 Konrad Lorenz: Die Rückseite des Spiegels. Versuch einer Naturgeschichte des menschlichen Erkennens. München 1973.
12 Immanuel Kant. Theorie – Werkausgabe Suhrkamp in 12 Bänden, Band 5. Frankfurt/M. 1984, p. 124.
13 Gerhard Vollmer: Was können wir wissen? 2 volumes. Stuttgart 1985.
14 Rupert Riedl: Evolution und Erkenntnis. Antworten auf Fragen aus unserer Zeit. München 1987.
15 Franz M. Wuketits: Biologie und Kausalität. Biologische Ansätze zur Kausalität, Determination und Freiheit. Berlin 1981.
16 E. Schierenberg, R. Cassada: Der Nematode Caenorhabditis elegans: ein entwicklungsbiologischer Modellorganismus. Biologie in unserer Zeit *16* (1986) pp. 1–7.

17 P. Brix: G.C. Lichtenberg. Der Physiker: Altes und Neues. Physikalische Blätter *41* (1985) No. 6, pp. 141–146.
18 Ilya Prigogine: From Being to Becoming: Time and Complexity in the Physical Sciences, W.C. Freeman and Company, San Francisco 1980.
19 Ilya Prigogine: Zeit, Struktur und Fluktuation (Nobel- Vortrag). Angew. Chemie *90* (1978) pp. 704–715.
20 P.R. Halmoss: Lectures on Ergodic Theory. Mathematical Society of Japan 1965, p. 9.
21 J.L. Lebowitz, O. Penrose: Modern Ergodic Theory. Phys. Today *23* (Feb. 1973) p. 29.
22 V.B. Havez: Ergodic Problems of Statistical Mechanics. New York 1968.
23 Taken from: The Norton Anthology of American Literature, Vol. 1, 2nd ed., W.W. Norton and Company, New York, London, 1985, p. 2474.

5. Mathematical and Physical Models of Deterministic Chaos

1 The original Wittgenstein quotes are taken from Ludwig Wittgenstein: Tractatus logico-philosophicus, edition suhrkamp. Frankfurt/M. 1971; L.W., Vermischte Bemerkungen, Bibliothek Suhrkamp, Band 535 (1977); L.W., Über Gewißheit, suhrkamp taschenbuch Wissenschaft 508. Frankfurt/M. 1984.
2 Ilya Prigogine: From Being to Becoming: Time and Complexity in the Physical Sciences, W.C. Freeman and Company, San Francisco 1980.
3 Henri Poincaré: Les Méthodes Nouvelles de la Mécanique Céleste. Paris 1892; English: Nasa Translation TTF-450–452. U.S. Federal Clearing House. Springfield/USA 1967.
4 E.N. Lorenz: Deterministic Nonperiodic Flow. J. Atmos. Sci. *20* (1963) p. 130.
5 H.G. Schuster: Deterministic Chaos. Weinheim 1984.
6 E.C. Zehmann: Catastrophe Theory. Scientific American (April 1970) pp. 63–83.
7 Heinz O. Peitgen, Peter H. Richter: Harmonie in Chaos und Kosmos. Pamphlet of the Städtische Sparkasse in Bremen 1984.
8 J. Moser: Stable and Random Motions in Dynamical Systems. Princeton University Press 1973; V.I. Arnold: Mathematical Methods of Classical Mechanics. Heidelberg/New York 1978.
9 R. Marcialis, R. Greenberg: Warming of Miranda during chaotic rotation. Nature *328* (1987) pp. 227–229.
10 B.B. Mandelbrot: The Fractal Geometry of Nature. San Francisco 1982.
11 J. Brickmann, H.-J. Bär: Chaos und fraktale Dimension. Nachrichten aus Chemie, Technik und Laboratorium, *34* (1986) pp. 566–572.
12 P. Pfeifer: Katalysatoroberflächen. Makromoleküle und Kolloid-Aggregate: Fraktale Dimension als versteckte Symmetrie unregelmäßiger Strukturen. Chimia *39* (1985) pp. 120–134.
13 Taken from: Wallace Stevens: The Palm at the End of the Mind, Selected Poems and a Play, Holly Stevens (Ed.), Vintage Books, New York 1972; pp. 166–168.

6. The World is Harmonic

1 Heinrich von Kleist: Sämtliche Werke, pp. 1134–1142. Leipzig 1930.
2 Johannes Kepler: Mysterium Cosmographicum.

3 Here I largely follow, in part verbatim, the discussion of Heinz O. Peitgen and Peter H. Richter in: The Beauty of Fractals. Images of Complex Dynamical Systems. Heidelberg/New York/Tokyo 1986.
4 B. B. Mandelbrot: Towards a Second Stage of Indeterminism in Science. Interdisc. Science Rev. *12* (1987) pp. 117–127.
5 B. B. Mandelbrot: Les Objects Fractals. Paris 1975; B. B. Mandelbrot: The Fractal Geometry of Nature. San Francisco 1982.
6 B. B. Mandelbrot: Fractals and the Rebirth of Iteration Theory, in: Heinz O. Peitgen, Peter H. Richter: The Beauty of Fractals (see Ref. 3).
7 Franz Xaver Pfeifer: Der Goldene Schnitt und dessen Erscheinungsformen in Mathematik, Natur und Kunst. Augsburg 1885.
8 P. H. Richter, R. Schranner: Leaf Arrangement-Geometry, Morphogenesis and Classification. Naturwiss. *65* (1978) pp. 319–327.
9 Johann Wolfgang von Goethe: Naturwissenschaftliche Schriften. Über die Spiraltendenz in der Natur. Goethe-dtv-Gesamtausgabe, Volume 39, p. 123 ff.
10 H. Meinhardt, A. Gierer: Applications of a Theory of Biological Pattern Formation based on Lateral Inhibition. J. Cell Sci. *15* (1974) pp. 321–346; A. Gierer, H. Meinhardt: A Theory of Biological Pattern Formation. Kybernetik *12* (1972) pp. 30–39.
11 Manfred Schroeder: Number Theory in Science and Communication. Berlin/Heidelberg/New York 1984.
12 Translated by Frederick Franck: The Book of Angelus Silesius, Knopf, New York 1976, p. 66.
13 Uwe H. Peters: Hölderlin. Wider die These vom edlen Simulanten. Reinbek 1982.
14 Translated by Michael Hamburger: Friedrich Hölderlin, Poems and Fragments, Cambridge University Press, Cambridge 1980.
15 K. M. Sullivan, D. M. Lilley: A Dominant Influence of Flanking Sequences on a Local Structural Transition in DNA. Cell *47* (1986) pp. 817–827.
16 Translated by Michael Hamburger: Poems of Celan, Persea Books, New York 1988.

7. Big Bang – Idea or Matter?

1 Werner Heisenberg: Der Teil und das Ganze. Gespräche im Umkreis der Atomphysik. München 1986.
2 Steven Weinberg: The First Three Minutes. A Modern View of the Origin of the Universe. Basic Books, New York, 1977.
3 Erich Jantsch: Die Selbstorganisation des Universums. München 1982.
4 Charles Darwin: On the Origin of Species by Means of Natural Selection or the Preservation of Favored Races in the Struggle for Life, John Murray, London 1859.
5 Journal of John Conduitt in R. S. Westfall: Never at Rest. Cambridge University Press 1980.
6 Thomas Kuhn: The Structure of Scientific Revolutions, University of Chicago Press, Chicago 1962.
7 Robert Spaemann, Reinhard Löw: Die Frage Wozu? Geschichte und Wiederentdeckung des teleologischen Denkens. München 1981.
8 Ludwig Wittgenstein: Tractatus logico-philosophicus. Frankfurt 1971.

9 Friedrich Engels: Ludwig Feuerbach und der Ausgang der klassischen Philosophie (1988). Berlin 1971.
10 Carsten Bresch: Zwischenstufe Leben. München 1977.
11 Friedrich Cramer: Die Evolution frißt ihre Kinder. Der Unterschied zwischen Newtonschen Bahnen und lebenden Wesen. Universitas *41* (1986) pp 1149–1156.
12 Ilya Prigogine, Isabelle Stengers: Order out of Chaos. Man's Dialogue with Nature, London 1984, p. 257ff.
13 Friedrich Cramer: Fundamental Complexity. A Concept in Biological Sciences and Beyond. Interdisciplinary Science Reviews *4* (1979) pp. 132–139.
14 Réné Descartes: Discours de la méthode pour bien conduire sa raison et chercher la vérité dans les sciences. Paris 1637.
15 Alfred Gierer: Die Physik, das Leben und die Seele. München 1986.
16 H.C. Schaller, H. Bodenmüller: Role of the neuropeptide head activator for nerve function and development. Biol. Chem. Hoppe-Seyler *366* (November 1985) pp. 1003–1007.
17 C. Nüsslein-Vollhardt, E. Wieschaus: Mutations affecting segment number and polarity in Drosophila. Nature *287* (1980) pp. 795–801.
18 W. Gehring, R. Nöthiger: The imaginal discs of Drosophila. In Developmental Systems: Insects, eds. S. Counce, C.H. Waddington, Vol. 2. New York 1973, pp. 211–290.
19 Rupert Sheldrake: Das schöpferische Universum. München 1984.
20 Réné Thom: Structural Stability and Morphogenesis. Translated by D.H. Fowler, W.A. Benjamin, Reading, Massachusetts 1975, p. 151f.
21 Erich Jantsch: Die Selbstorganisation des Universums. München 1982, p. 34f.
22 Ilya Prigogine, Isabelle Stengers: Order out of Chaos. Man's dialogue with Nature, London 1984, p. 14.
23 Alister Hardy: Der Mensch das betende Tier. Religiosität als Faktor der Evolution. Stuttgart 1979.
24 Hoimar v. Ditfurth: Wir sind nicht nur von dieser Welt. München 1981.
25 Reinhard Löw: Neue Träume eines Geistersehers. Bemerkungen zu Hoimar von Ditfurth. Scheidewege *12* (1982) pp. 687–697.
26 Roald Hoffmann. The Metamict State. University Press of Florida, p. 77.

8. Aging and Death – Our Time

1 Taken from Plato's Phaedo in J.D. Kaplan (Ed.), Dialogues of Plato, Pocket Books, Inc., New York 1950.
2 Ilya Prigogine: From Being to Becoming: Time and Complexity in the Physical Sciences, W.C. Freeman and Company, San Francisco 1980.
3 Denis Diderot: D'Alamberts Dream, "Conversation Between d'Alambert and Diderot", Hammondsworth, Engl. Penguin Books, 1976, pp. 158–159.
4 Ludwig Boltzmann: Über die Unentbehrlichkeit der Atomistik in der Naturwissenschaft. Annalen der Physik und Chemie *396* (1987) pp. 232–247.
5 Karl Popper: Ausgangspunkte. Hamburg 1979, p. 233.
6 Here I largely follow, in part verbatim, the discussion of Bernd O. Küppers: Entropie. Evolution und Zeitstruktur. Futura *4* (1986), p. 19.
7 Carl Friedrich v. Weizsäcker: Einheit der Natur. München 1971.
8 Michael Drieschner: Einführung in die Naturphilosophie. Darmstadt 1981.

9 R. Jost: Erinnerungen: Erlesenes und Erlebtes. Physikalische Blätter *40* (1984) pp. 178–181.

10 Konrad Lorenz: Kants Lehre vom Apriorischen im Lichte gegenwärtiger Biologie. Blätter für Deutsche Philosophie *15* (1941), pp. 94–125.

11 Georg Picht: Die Zeit und die Modalitäten, in: H.P. Dürr (Ed.), Quanten und Felder. Braunschweig 1971.

12 Ecclesiastes, 3, 1–8, King James Bible.

13 H.C. Schröder, R. Messer, H.J. Breter, W.E.G. Müller: Evidence for age-dependent impairment of ovalbumin heterogeneous nuclear RNA (hnRNA) processing in hen oviduct. Mech. Ageing Dev. *30* (1985) p. 319.

14 Friedrich Cramer: Death – from Microscopic to Macroscopic Disorder, in: E. Frehland (Ed.), Synergetics – from Microscopic to Macroscopic Order. Heidelberg/Berlin 1984, pp. 220–228.

15 Heinz C. Schröder: Biochemische Grundlagen des Alterns. Chemie in unserer Zeit *20* (1986) pp. 128–138.

16 Leslie E. Orgel: The Maintenance of Accuracy of Protein Synthesis and its Relevance to Ageing. Proc. Natl. Acad. Sci. USA *49* (1963) pp. 517–521; Leslie E. Orgel: The Maintenance of Accuracy of Protein Synthesis and its Relevance to Ageing, a Correction. Proc. Natl. Acad. Sci. USA *67* (1970) p. 1476.

17 Friedrich Cramer, Wolfang Freist: Molecular Recognition by Energy Dissipation, a New Enzymatic Principle: The Example Isoleucine-Valine. Accounts of Chemical Research *20* (1987) pp. 79–84.

18 U. Englisch, D. Gauss, W. Freist, S. Englisch, H. Sternbach, F.v.d. Haar: Error Rates of the Replication and Expression of Genetic Information. Angew. Chem. Int. Ed. Engl. *24* (1985) pp. 1015–1025; Angew. Chemie *97* (1985) pp. 1033–1043.

19 H.-J. Gabius, S. Gabius, G. Graupner, F. Cramer, S. Rehm: Aminoacyl-tRNA Synthetases in Liver, Spleen and Small Intestine of Aged Leukemic and Aged Normal Mice. Z. Natf. *38c* (1983) pp. 881–882; H.-J. Gabius, S. Goldbach, G. Graupner, S. Rehm, F. Cramer: Organ Pattern of Age-related Changes in the Aminoacyl-tRNA Synthetases Activities of the Mouse. Mechan. Ageing Develop. *20* (1982) pp. 305–313; H.-J. Gabius, G. Graupner, F. Cramer: Activity Pattern of Aminoacyl-tRNA Synthetases, tRNA Methylases, Arginyltransferase and Tubulin: Tyrosine Ligase during Development and Ageing of Caenorhabditis elegans. Eur. J. Biochem. *131* (1983) pp. 231–234; H.-J. Gabius, R. Engelhardt, F. Deerberg, F. Cramer: Age-Related Changes in Different Steps of Protein Synthesis of Liver and Kidney of Rats. FEBS Lett. *160* (1983) pp. 115–118.

20 A. Garcia-Tejedor, F. Moran, F. Montero: Influence of the Hypercyclic Organization on the Error Threshold. J. Theor. Biol. *127* (1987) pp. 393–402.

21 Seneca: De brevitate vitae. Translated by Moses Hadas in The Stoic Philosophy of Seneca, W.W. Norton and Company, New York 1958.

22 Wilhelm Doerr: Altern – Schicksal oder Krankheit. Sitzungsbericht 83/4 der Heidelberger Akademie der Wissenschaften, Heidelberg 1983.

23 Martin Gregor-Dellin: Alt werden, heißt, selbst ein neues Geschäft antreten. Frankfurter Allgemeine Zeitung No. 103 of 4 May 1985.

24 Taken from The Norton Anthology of Modern Poetry, 2nd ed., Richard Ellmann, Robert O'Clair (Eds.), W.W. Norton and Company, New York, London, 1988, p. 161.

9. Fundamental Complexity – Intrinsic Limitations

1 N. Pippenger: Complexity Theory. Scientific American (June 1978) pp. 90–100.
2 Friedrich Cramer: Fundamental Complexity. A Concept in Biological Sciences and Beyond. Interdisciplinary Science Reviews *4* (1979) pp. 132–139.
3 G. J. Chaitin: Randomness and Mathematical Proof. Scientific American (May 1975) pp. 47–52.
4 Friedrich A. v. Hayek: Die Theorie komplexer Phänomene. Tübingen 1972.
5 Friedrich Cramer: Fortschritt durch Verzicht. München 1975, p. 56f.
6 Konrad Lorenz: Die Rückseite des Spiegels. Versuch einer Naturgeschichte menschlichen Erkennens. München 1975.
7 Karl Popper: Das Elend des Historizismus. Tübingen 1974.
8 Manfred Eigen, Ruthild Winkler: Das Spiel. München 1975.
9 Pascual Jordan: Begegnungen. Oldenburg 1971, pp. 97–103.
10 Based on a lecture delivered at the University of Basel on 5 February 1982.
11 G. W. F. Hegel: Werke, Vol. 19 (p. 111) and Vol. 20 (p. 462). Frankfurt 1986. The translation of the quote from Hegel's lectures is taken from: Hegel's Lectures on the History of Philosophy, Vol. 3, translated by E. S. Haldane and Frances H. Simson, The Humanities Press Inc., New York, 1955, p. 553.
12 Jürgen Habermas, Niklas Luhmann: Theorie der Gesellschaft oder Sozialtechnologie. Frankfurt/Main 1974, p. 148ff.
13 Heinrich v. Kleist: Über die allmähliche Verfertigung der Gedanken beim Reden. Sämtliche Werke und Briefe. München 1977, p. 321.
14 Hubert Markl: Natur als Kulturaufgabe. Über die Beziehung des Menschen zur lebendigen Natur. Stuttgart 1986.
15 Hans Jonas: Das Prinzip Verantwortung. Frankfurt/Main 1979.
16 Kurt Gödel: Über formal-unentscheidbare Sätze der Principia Mathematica und verwandter Systeme I. Monatshefte für Mathematik und Physik *38* (1931) pp. 173–198.
17 Douglas R. Hofstaedter: Gödel, Escher, Bach. An Eternal Braid, Vintage Books, 1980.
18 Odo Marquardt: Abschied vom Prinzipiellen. Philosophische Studien. Stuttgart 1981, p. 16f.
19 Friedrich A. v. Hayek: Die drei Quellen der menschlichen Werte. Tübingen 1979.
20 Niklas Luhmann: Vertrauen. Ein Mechanismus der Reduktion von sozialer Komplexität. Stuttgart[2] 1973.
21 Igor R. Schafarewitsch: Über einige Tendenzen in der Entwicklung der Mathematik. Jahrbuch der Akademie der Wissenschaften in Göttingen. Göttingen 1973, pp. 31–36.
22 The First Epistle of Paul the Apostle to the Corinthians, 13, 12, King James Bible. 23 Immanuel Kant: Kritik der reinen Vernunft. Theorie- Werkausgabe Suhrkamp in 12 volumes, Volume 5. Wiesbaden 1956. pp. 267/268.
24 Translated by Michael Hamburger: Friedrich Hölderlin, Poems and Fragments, Cambridge University Press, Cambridge 1980.

Index

Page numbers in italics refer to Figures.

Sources

We are indebted to the following persons and publishers for permission to reprint figures and extracts of texts:

Aero Service Corp., Philadelphia, MA (USA)
Alfred A. Knopf, New York, NY (USA)
Cambridge University Press, Cambridge (United Kingdom)
Dr. Bruce Alberts, San Francisco, CA (USA)
Dr. David Epel, Pacific Grove, CA (USA)
Dr. Jeanette Hudson
Fritz Schwörer and coworkers, Max-Planck-Institut für Strahlenchemie, Mülheim/Ruhr (Germany)
Garland Publishing Inc., New York, NY (USA)
John Murray Publishers, London (United Kingdom)
Macmillan Academic and Professional, London (United Kingdom)
McCrone and Associates, Chicago, IL (USA)
Mechthild Ziemer, Göttingen (Germany)
Michael Hamburger, Middleton (United Kingdom)
Oxford University Press, Oxford (United Kingdom)
Persea Books, New York, NY (USA)
Prof. Roald Hoffmann, Ithaca, NY (USA)
Schweizerische Gewitterstation, Lugano (Switzerland)
Simon & Schuster, New York, NY (USA)
Springer Verlag, Heidelberg (Germany)
The Quaestor and Factor, University of St. Andrews, Fife (United Kingdom)
Twayne Publishers, Boston, MA (USA)
University Presses of Florida, Gainesville, FL (USA)
Vintage Books, New York, NY (USA)
W.A. Benjamin, Reading, MA (USA)
W.W. Norton & Co., London (United Kingdom)
Peters Fraser & Dunlop Group Ltd. (UK)